Holonomy and Parallel Spinors in Lorentzian Geometry

DISSERTATION

zur Erlangung des akademischen Grades
doctor rerum naturalium
(Dr. rer. nat.)
im Fach Mathematik

eingereicht an der
Mathematisch-Naturwissenschaftlichen Fakultät II
der Humboldt-Universität zu Berlin

von
Thomas Leistner

Präsident der Humboldt-Universität zu Berlin:
Prof. Dr. Jürgen Mlynek

Dekan der Mathematisch-Naturwissenschaftlichen Fakultät II:
Prof. Dr. Elmar Kulke

Gutachter:

1. Prof. Dr. H. Baum (Humboldt-Universität zu Berlin)
2. Prof. Dr. D. Alekseevsky (University of Hull, UK)
3. Prof. Dr. L. Schwachhöfer (Universität Dortmund)

eingereicht am: 20. Mai 2003
Tag der mündlichen Prüfung: 25. September 2003

Bibliografische Information Der Deutschen Bibliothek

Die Deutsche Bibliothek verzeichnet diese Publikation in der Deutschen Nationalbibliografie; detaillierte bibliografische Daten sind im Internet über http://dnb.ddb.de abrufbar.

ISBN 3-8325-0472-9

Logos Verlag Berlin
Comeniushof, Gubener Str. 47,
10243 Berlin
Tel.: +49 030 42 85 10 90
Fax: +49 030 42 85 10 92
INTERNET: http://www.logos-verlag.de

Abstract

In this thesis, Lorentzian manifolds and their holonomy groups are studied under the condition that they admit parallel spinors.
In the first chapter the basic concepts are introduced. We show that a semi-Riemannian product manifold admits a parallel spinor if and only if each of the factors admits a parallel spinor. Using the Wu-Decomposition theorem and the vector field associated to the parallel spinor, we can solve the existence problem for parallel spinors for Lorentzian manifolds, which decompose completely into irreducible factors.
The remaining indecomposable, non-irreducible Lorentzian manifolds are treated in the second chapter. Their holonomy group is contained in the parabolic group $(\mathbb{R}^* \times SO(n)) \ltimes \mathbb{R}^n$. We describe algebraic properties of such subgroups. We show that the $SO(n)$–component of the holonomy group is the holonomy group of the screen-bundle. Furthermore we give sufficient conditions on the local form of the manifolds for the property, that the $SO(n)$–component is a Riemannian holonomy group.
In the third chapter, indecomposable, non-irreducible Lorentzian manifolds are studied, assuming that they carry parallel spinors. This property implies that the $\mathbb{R}^*$–component of the holonomy has to be trivial and the $SO(n)$–component cannot be abelian. The existence of parallel spinors only depends on the $SO(n)$–component. If the latter contains $SU(n/2)$, then the existence of a parallel spinor implies that it is equal to $SU(n/2)$. Furthermore, the sufficient conditions of chapter two, now for the existence of a parallel spinor, can be weakened. Thus one gets a large class of examples for Lorentzian manifolds with parallel spinors. For a special class of examples, we show that an indecomposable Lorentzian manifold has abelian holonomy if and only if it is a *pp*–wave.
In the last chapter we study the question, which Lie groups may occur as $SO(n)$–component of an indecomposable, non-irreducible Lorentzian manifold. We formulate an algebraic criterion, based on the first Bianchi identity. Using this criterion we get the following result: If G is the $SO(n)$–component of an indecomposable, non-irreducible Lorentzian manifold, then it is a Riemannian holonomy group, provided that it acts irreducibly and is simple, or provided that $G \subset U(n/2)$. This implies that an indecomposable Lorentzian manifold with parallel spinor field does not have a holonomy group of coupled type.

Danksagung Vor allem möchte ich meiner Mentorin Prof. Helga Baum für die Betreuung, Unterstützung und für wertvolle Hinweise während der Arbeit an dem Thema danken. Weiterer Dank gilt Thomas Neukirchner, der immer Zeit für ein Gespräch hatte und jede Menge Anregeung bereithielt.
Danken möchte ich meinen Eltern, für Unterstützung in jeder Hinsicht, und natürlich Rea und Lena, ohne die alles viel weniger Spaß gemacht hätte.

Was er sah, war sinnverwirrend. In einer krausen, kindlich dick aufgetragenen Schrift, die Imma Spoelmanns besondere Federhaltung erkennen ließ, bedeckte ein phantastischer Hokuspokus, ein Hexensabbat verschränkter Runen die Seiten. Griechische Schriftzeichen waren mit lateinischen und mit Ziffern in verschiedener Höhe verkoppelt, mit Kreuzen und Strichen durchsetzt, ober- und unterhalb waagerechter Linien zeltartig überdacht, durch Doppelstrichelchen gleichgewertet, durch runde Klammern zusammengefaßt, durch eckige Klammern zu großen Formelmassen vereinigt. Einzelne Buchstaben, wie Schildwachen vorgeschoben, waren rechts oberhalb der umklammerten Gruppen ausgesetzt. Kabbalistische Male, vollständig unverständlich dem Laiensinn, umfaßten mit ihren Armen Buchstaben und Zahlen, während Zahlenbrüche ihnen voranstanden und Zahlen und Buchstaben ihnen zu Häupten und Füßen schwebten. Sonderbare Silben, Abkürzungen geheimnisvoller Worte waren überall eingestreut, und zwischen den nekromantischen Kolonnen standen geschriebene Sätze und Bemerkungen in täglicher Sprache, deren Sinn gleichwohl so hoch über allen menschlichen Dingen war, daß man sie lesen konnte, ohne mehr davon zu verstehen als von einem Zaubergemurmel.

„Und über diesen gottlosen Künsten wollen Sie den schönen Vormittag versäumen? ... Nein“, rief er, „heute dürfen Sie keine Algebra treiben, Fräulein Imma, oder im luftleeren Raume spielen, wie Sie es nennen! Sehen Sie doch die Sonne! ... “

Thomas Mann, Königliche Hoheit

Contents

Introduction 1

1 Holonomy and parallel spinors 15

1.1 Holonomy - Definition and basic facts 15

1.1.1 Holonomy of a connections in a principal fibre bundle 15

1.1.2 Holonomy and parallel sections in vector bundles 17

1.1.3 Local and infinitesimal holonomy group 18

1.1.4 Holonomy of semi-Riemannian manifolds 20

1.2 Parallel Spinors on semi-Riemannian manifolds 22

1.2.1 Spinor representations and the spinor bundle 22

1.2.2 First properties of parallel spinors 25

1.2.3 Parallel spinors on product manifolds 26

1.3 Decomposition of a Lorentzian manifold with parallel spinors 31

2 Indecomposable Lorentzian manifolds 35

2.1 The holonomy of an indecomposable Lorentzian manifold 35

2.1.1 Basic properties 35

2.1.2 The $\mathfrak{so}(n)$-projection as holonomy group 38

2.1.3 Results of A. Ikemakhen and L. Berard-Bergery 41

2.1.4 14 holonomy types of space-time 44

2.2 Local description of an indecomposable Lorentzian manifold 45

2.2.1 Parallel distributions and foliations 45

2.2.2 Local coordinates 48

2.3 Walker coordinates and holonomy 51

2.3.1 The projected connection 52

2.3.2 Families of Riemannian metrics 56

2.3.3 Holonomy with Riemannian holonomy as $SO(n)$–component . . 59

3 Parallel spinors on indecomposable Lorentzian manifolds **63**
3.1 Parallel spinors and indecomposable holonomy 64
3.1.1 Notations 64
3.1.2 Basic algebraic properties 64
3.1.3 Results in low dimensions 67
3.1.4 Holonomy containing $\mathfrak{su}(n)$ 69
3.1.5 Further algebraic consequences 71
3.2 Conditions in local coordinates and examples 72
3.2.1 The local situation 72
3.2.2 Sufficient conditions for families of G–metrics admitting parallel spinors 74
3.2.3 Brinkmann-waves with abelian holonomy 76
3.2.4 The Ricci-isotropy 77

4 Weak-Berger algebras and Lorentzian holonomy **79**
4.1 Berger algebras, weak-Berger algebras and Lorentzian holonomy 80
4.1.1 Berger and weak-Berger algebras 80
4.1.2 Lorentzian holonomy and weak-Berger algebras 83
4.1.3 Real and complex weak Berger algebras 85
4.2 Weak-Berger algebras of real type 86
4.2.1 Irreducible, complex, orthogonal, semisimple Lie algebras 88
4.2.2 Irreducible complex weak-Berger algebras 89
4.2.3 Berger algebras, weak Berger algebras and spanning triples . . . 93
4.2.4 Properties of root systems 95
4.3 Simple complex weak-Berger algebras of real type 99
4.3.1 Representations with roots as highest weight 99
4.3.2 Representations with planar spanning triples 103
4.3.3 Representations with the property (SII) and weight zero 107
4.3.4 Representations with the property (SII) where zero is no weight 112
4.3.5 Consequences for simple weak-Berger algebras of real type 128
4.4 Weak-Berger algebras of non-real type 130
4.4.1 The first prolongation of a Lie algebra of non-real type 131
4.4.2 Consequences for Berger and weak-Berger algebras 135
4.4.3 Lie algebras with non-trivial first prolongation 137
4.4.4 The result and consequences 138

A Complex semi-simple Lie algebras and their representations **141**
A.1 Structure of a complex semi-simple Lie algebra 141
A.2 Basic facts about complex representations 144

B Representations of real Lie algebras **147**
B.1 Preliminaries 147
B.2 Irreducible complex representations of real Lie algebras 149
B.3 Irreducible real representations 150
B.3.1 Representations of real type 150
B.3.2 Representations of non-real type 151
B.3.3 Weights of real type and non-real type representations 153
B.4 Orthogonal real representations 154
B.4.1 Orthogonal and symplectic representations of real type 155
B.4.2 Orthogonal representations of non-real type 156

C Algebras with first prolongation V^* **161**
C.1 $\mathfrak{co}(n,\mathbb{C})$ acting on$V := \mathbb{C}^n$ 161
C.2 $\mathfrak{gl}(n,\mathbb{C})$ acting on $V := \odot^2\mathbb{C}^n$ 163
C.3 $\mathfrak{gl}(n,\mathbb{C})$ acting on $V := \wedge^2\mathbb{C}^n$ 165
C.4 $\mathfrak{sl}(\mathfrak{gl}(n,\mathbb{C}) \oplus \mathfrak{gl}(m,\mathbb{C}))$ acting on $V := \mathbb{C}^n \otimes \mathbb{C}^m$ 167
C.5 $\mathbb{C}\, Id \oplus \mathfrak{spin}(10,\mathbb{C})$ and $\mathbb{C}\, Id \oplus \mathfrak{e}_6$ 168
C.5.1 The Lie algebras and groups of type E_8, E_7 and E_6 168
C.5.2 Decomposition of $\mathfrak{e}_8$ 170
C.5.3 The exceptional Kählerian symmetric spaces 172

Bibliography **173**

Introduction

Parallelism is one of the basic concepts in differential geometry. Its definition needs a differentiable manifold as well as a *covariant derivative* on the tangent spaces, usually denoted by ∇. It generalizes the usual derivative in a vector space and defines a *parallel displacement* along piecewise smooth curves. For a curve in the manifold, starting at the point p, and a vector V_p in the tangent space of p there is a uniquely determined vector field along the curve, which solves the following initial value problem: its covariant derivative in direction to the curve is zero and its value in p is V_p. Thus ∇ defines a parallel displacement along this curve assigning to V_p the value of this vector field at a point of the curve. The parallel displacement along a curve gives an isomorphism between the tangent spaces in the start- and the endpoint of the curve.

The parallel displacement can be used to obtain important geometric objects on the manifold. For example, geodesics which are generalizations of straight lines are defined as curves whose tangent vector is parallel displaced along the curve.
Since ∇ links tangent spaces in different points via the parallel displacement, it bears also the name *connection*. But it may also be useful to consider the parallel displacements along piecewise smooth curves starting and ending at the same point p (*loops* around p). Then it is an automorphism of T_pM. For a flat space it is the identity, of course, but for a curved manifold this is not always the case. For the sphere any rotation of T_pM can be obtained by parallel displacements, as the picture should suggest.

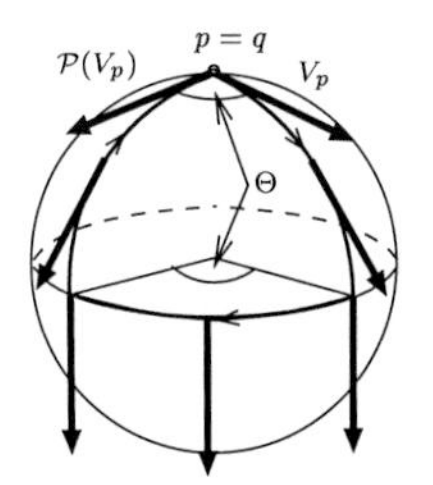

$\mathcal{P}(V_p)$ is the parallel displacement of V_p along the loop .

This allows it to define the *holonomy group* in a point $p \in M$, as the set of parallel displacements along loops around p. This set forms a Lie group which is given together with its representation on the tangent space of p. Thats why one also refers to it as holonomy representation. It is contained in the automorphism group of T_pM. In the picture it is the rotation group of the tangent plane in p.

The holonomy group is an algebraic object, which of course can be studied by methods of algebra and its powerful tools. This yields to a double edged situation: geometric facts *can* be proven by algebraic methods, but the absence of a direct geometric proof often is considered as a lack.
The origin of the concept of holonomy lies in physics. In 1896 H. Hertz introduced the terms *holonomic* and *nonholonomic* constraints in a mechanical system. Holonomy groups where introduced by Cartan [Car23a], [Car23b], [Car26].
If the manifold we start with is a semi-Riemannian manifold, i.e. equipped with a metric, then the tangent spaces are inner product vector spaces. In this situation exists a uniquely defined connection which is compatible with this metric and torsionfree. It is called *Levi-Civita connection* and also denoted by ∇. For a semi-Riemannian manifold with metric h of signature (r, s) the holonomy group of the Levi-Civita connection is contained in $O(T_pM, h) \simeq O(r, s)$.

The covariant derivative of a manifold also characterizes directions which are "flat". These are directions of *parallel vector fields*, i.e. vector fields whose covariant derivative is zero in *any* direction. The covariant derivative is defined on the tangent bundle but it can be extended to a covariant derivative in every geometric vector bundle such as the dual tangent bundle, tensor bundles of higher order or the spinor bundle. It leads to the concept of *parallel sections* of such a bundle.
As the holonomy groups acts on a tangent space — which is of course a fibre of the tangent bundle — it acts also on the fibres of any geometric vector bundle. For a geometric vector bundle $\pi : E \to M$ with covariant derivative ∇ and corresponding holonomy group $Hol_p(\nabla)$ in a point p, there is the following fundamental relation between parallel sections and fixed vectors under holonomy representation:

$$\left\{ \begin{array}{l} \text{Sections } \sigma \text{ of } E \text{ which are} \\ \text{parallel w.r.t. } \nabla, \text{ i.e. } \sigma \in \\ C^\infty(M, E) \text{ with } \pi \circ \sigma = \pi \\ \text{and } \nabla\sigma = 0 \end{array} \right\} \overset{\sim}{\longleftrightarrow} \left\{ \begin{array}{l} \text{Vectors } V_p \text{ in the fibre of } E \text{ over} \\ p \text{ which are fixed under the holo-} \\ \text{nomy group in } p, \text{ i.e. } \mathcal{P}(V_p) = V_p \\ \text{for all } \mathcal{P} \in Hol_p(\nabla) \end{array} \right\} \quad (1)$$

Thus algebraic properties of the holonomy representation, such as invariant vectors, endomorphisms or forms, correspond to parallel sections in geometric vector bundles, and these define geometric structures of the manifold. The geometric problem of finding such structures can be transposed to the algebraic problem to find fixed vectors, i.e. trivial subrepresentations, of the corresponding representation of the holonomy group.

Beside holonomy groups — to which we will return later on in this introduction — the spinor bundle will play a role in the present thesis. The spinor bundle can be defined for semi-Riemannian manifolds which are spin. That is to say the bundle of orthonormal frames admits a reduction of the twofold covering of the orthogonal group by the spin group. It is a (complex) vector bundle associated to this reduction and to the (complex) spin representation of the spin group. It has its origin in physics too. In 1928 P.A.M. Dirac introduced spinors to formulate the wave equation for the electron in the for 4–dimensional Minkowski space-time, now called Dirac equation. Another important spinor field equation has its origin in physics. The twistor equation was introduced by R. Penrose in the theory of general relativity. But spinors also became important in mathematics considering the following spinor field equations:

$$\begin{aligned} \nabla\varphi &= 0 \quad \text{equation for a parallel spinor,} \\ \nabla_X\varphi - \lambda X\cdot\varphi &= 0 \quad \text{Killing equation } (\lambda\in\mathbb{C}), \\ D\varphi &= 0 \quad \text{Dirac equation,} \\ \nabla_X\varphi - \tfrac{1}{n}X\cdot D\varphi &= 0 \quad \text{twistor equation.} \end{aligned}$$

Parallel spinors are special solutions of the Dirac and the twistor equation.
In spite of its origin in special and general relativity, where a 4–dimensional Lorentzian manifold serves as the model of space and time, these equations where studied first of all in Riemannian geometry, i.e. for manifolds with a positive definite metric tensor. A major result was the Atiyah-Singer index theorem around 1960 which coupled the dimensions of the kernel and the image of the Dirac operator to the topology of the manifold. Investigations of the eigenvalues of the Dirac operator leads to the Killing equation. T. Friedrich [Fri80] showed that the limit case in the eigenvalue estimate on compact Riemannian spin manifolds with non-negative scalar curvature is realized by a Killing spinor. Furthermore one was interested in the classification of Riemannian manifolds admitting parallel, Killing or twistor spinors. The classification of holonomy groups of Riemannian manifolds with parallel spinors was obtained by M. Wang [Wan89] using the classification of Riemannian holonomy groups — see below — and was used to describe the geometric structures of Riemannian manifolds with Killing spinors. C. Bär [Bär93] showed that a simply connected Riemannian manifold admits a Killing spinor with real λ if and only if its cone is a Riemannian manifold with parallel spinor.
Again developments in physics caused that spin geometry became important in Lorentz geometry anew, also in dimensions higher than 4. Supergravity and superstring theories use dimensions up to eleven and are considered as generalizations of general relativity. Here a solution of a modified Killing equation on a Lorentzian manifold describes a supersymmetry. In some situations this equation degenerates to the equation of a parallel spinor. All in all the question for Lorentzian manifolds with parallel, Killing or twistor spinors became of interest again.
Some references on Killing and twistor spinors in the Lorentzian as well as in general pseudo-Riemannian context are the following: [NW84], [DNP86] [FO00], [FOP02],

[FO02] ,[Boh03], ,[Lei01], [BL03], [Lei03], ,[Kat99], [Kat00], [Bau99b], [Bau99a], [Bau00b], [Bau00a], [Bau03], [BK03].

If one is interested in semi-Riemannian manifolds which carry parallel spinors, because of Principle (1) it is reasonable to ask the following questions:

(Q1) Which Lie groups can occur as holonomy of a semi-Riemannian manifold?

(Q2) Which of these have trivial spinor subrepresentations?

There is a main result in holonomy theory which create hope to solve problem (Q1). It is the following decomposition theorem:

A simply connected, complete semi-Riemannian manifold is globally isometric to a product of simply connected, complete semi-Riemannian manifolds with trivial or weakly irreducible holonomy representation.

This theorem was proved first for Riemannian manifolds by G. de Rham [dR52], and in the general case by H. Wu [Wu64]. For Riemannian manifolds without the completeness assumption A. Borel and A. Lichnerowicz [BL52] proved the above statement but only asserting a local isometry. Sometimes this property is called *Borel-Lichnerowicz property.*
Instead of "weakly irreducible" we will use throughout the whole thesis the term "indecomposable" from now on. An orthogonal representation is *indecomposable* if on every invariant subspace (non-trivial and proper) the scalar product degenerates.
Clearly if the scalar product is positive definite — as it is the case in the Riemannian setting — indecomposable is the same as irreducible.
The second major result in holonomy theory. is due to M. Berger. It is a list of all irreducibly acting Lie groups which can occur as holonomy group of a semi-Riemannian manifold ([Ber55] for non-symmetric, [Ber57] for symmetric semi-Riemannian manifolds). Further investigations exclude some groups of the Berger list in the Riemannian case ([Ale68], [BG72] and [Bry87]) as well as in the semi-Riemannian case (see [Bry99a]), add some groups, and first of all yield local examples for the remaining groups. Thus there is a complete list of irreducibly acting groups which *are* holonomy groups of simply connected semi-Riemannian manifolds.
The Berger list was achieved by means of the third main result in holonomy theory, which is the holonomy theorem of W. Ambrose and I.M. Singer [AS53]. It is formulated in a more general context than the metrical one. Roughly speaking it asserts that the holonomy algebra of a connection in a principle fibre bundle over a connected manifold is spanned by the curvature endomorphisms. It was independently proved for the case of analytic affine manifolds by A. Nijenhuis [Nij53]. Affine manifolds are manifolds with

a connection ∇. There is a result by J. Hano and H. Ozeki [HO65] that every closed group can be realized as holonomy group of a connection, possibly with torsion.
But if one poses conditions to the torsion it is not clear which holonomy groups may exist. For example asking for holonomy groups of connections with vanishing torsion yields a classification problem. For such a connection the first and second Bianchi identity are valid. By the Ambrose-Singer holonomy theorem these identities put algebraic constraints to the holonomy algebra. The first Bianchi identity gives the following criterion: If $\mathfrak{h}$ is the Lie algebra of the holonomy group of a torsionfree connection, acting on the vector space $V \simeq T_pM$, then it obeys

$$\mathfrak{h} = span\left\{R(u,v) \mid u,v \in V, R \in \mathcal{K}(\mathfrak{h})\right\}, \tag{2}$$

where

$$\mathcal{K}(\mathfrak{h}) \ := \ \left\{R \in \wedge^2 V^* \otimes \mathfrak{h} \mid R(u,v)w + R(v,w)u + R(w,u)v = 0 \text{ for all } u,v,w \in V\right\}$$

is the space of curvature endomorphisms. This is the well known *Berger criterion* and Lie algebras satisfying this criterion are usually called *Berger algebras* or *admissible*.
The second Bianchi identity gives a further criterion if the space is non-symmetric.
There is also a Berger list for affine manifolds with torsionfree connections. The exact classification of such holonomy groups is due to S. Merkulov and L. Schwachhöfer [MS99], see also [CMS95] and [Bry99a].
Turning back to the Riemannian situation one can state that the holonomy problem (Q1) is solved for simply connected complete manifolds. Every such holonomy group is a product of entries of the Berger list. This list splits in a short one for non-symmetric manifolds with the entries $SO(n)$, $U(n/2)$, $SU(n/2)$, $Sp(n/4)$, $Sp(1) \cdot Sp(n/4)$, G_2 and $Spin(7)$, and a rather long list of symmetric spaces (see for example [Hel78] or [Bes87]).
In the Riemannian setting, there is a shorter proof — less algebraic, more geometric — of the non-symmetric list by Simons [Sim62], who proved that non-symmetric Riemannian holonomy groups act transitively on the unit sphere.
These holonomy groups correspond to the following geometric structures on the manifolds:

$SO(n)$: This is the holonomy of a generically curved Riemannian manifold, there is no additional structure induced by the holonomy group.

$U(n)$: In this case there is a parallel hermitian structure. These manifolds are called Kähler manifolds. Their dimension is $2n$.

$SU(n)$: One has a parallel hermitian structure too and in addition these manifolds are Ricci-flat. Their dimension equals to $2n$ and they are called special Kähler manifolds.

$Sp(n)$: There is a parallel quaternionic structure. The dimension is $4n$. These manifolds are Ricci-flat and are called hyper-Kähler manifolds.

$Sp(1) \cdot Sp(n)$: There is a quaternionic-Kähler structure, i.e there is a parallel 3–dimensional subbundle in the endomorphism bundle, which is generated by three almost complex structures. These manifolds are $4n$–dimensional and they are called quaternionic-Kähler-manifolds.

G_2 and $Spin(7)$: These are the exceptional cases because they occur only in the dimensions 7 and 8. In case of a G_2–manifold there is a parallel three-form and in case of a $Spin(7)$–manifold a parallel four-form. Both manifolds are Ricci-flat.

In the Riemannian setting question (Q1) was extended to the question:

(Q3) Are there Riemannian manifolds with given holonomy group and prescribed topological properties, in particular, are there compact manifolds?

For complete examples for the exceptional holonomies we refer to R. Bryant and S. Salamon [BS89]. D. Joyce [Joy00] constructed compact examples in particular for the exceptional Riemannian holonomy groups G_2 and $Spin(7)$. About the non simply connected case not very much is known (see [Bry99a]).
Nevertheless for the Riemannian setting we are at a point where we can return to the question of parallel spinors on simply connected complete Riemannian manifolds. First one should to remark that, due to N. Hitchin [Hit74] the existence of a parallel spinor on a Riemannian manifold forces the Ricci-curvature to be zero. Since we are in the case of complete manifolds "locally symmetric" is the same as "symmetric", but symmetric Riemannian manifolds with zero Ricci-curvature are flat. Hence, in order to answer question (Q2) one has to check the short Berger list for trivial subrepresentations of the spin representation. This was done by M. Y. Wang [Wan89] and leads to the following list:

Wang list: Only the irreducible Riemannian manifolds with holonomy $SU(n)$, $Sp(n)$, G_2 and $Spin(7)$ and products of these admit parallel spinors.

In the general semi-Riemannian setting there is a result analogous to that of Wang by H. Baum and I. Kath [BK99]. They showed which of the *irreducible, non locally symmetric* semi-Riemannian holonomy groups have a trivial subrepresentation of its spin representation. Furthermore they calculated explicitely these trivial subrepresentations, and achieved results about the dimension and causal type of the space of parallel spinor fields.
For symmetric spaces there is an analogous result as in the Riemannian setting: If the transvection group of a semi-Riemannian symmetric space is semi-simple — in

particular if the space is irreducible — the Ricci-tensor has to be non-degenerate (see for example [Neu02]). On the other hand the existence of a parallel spinor implies a totally isotropic Ricci-endomorphism [Bau81], i.e. a degenerate Ricci-tensor. Hence an irreducible locally symmetric non-flat semi-Riemannian space cannot have a parallel spinor. Thus the problem (Q2) is solved for semi-Riemannian manifolds with *irreducible* holonomy representation by [BK99].
But for general semi-Riemannian manifolds the problem (Q1) as well as the problem (Q2) is open, because semi-Riemannian holonomy representations with degenerate invariant subspaces could not be treated. In other words, until now there is no *classification of indecomposable, non-irreducible Berger algebras.*
To describe the general situation more adequately we have to reconsider the Wu-decomposition. Decomposing the holonomy representation into parts, which cannot be decomposed further without degeneration of the metric, one achieves the decomposition into indecomposable manifolds. For those *indecomposable factors* can occur the following cases:

1. It is an irreducible semi-Riemannian holonomy representation of the Berger list.
2. It is a non-irreducible holonomy representation. Here one distinguishes:
   (a) The factor is not completely reducible, i.e. degenerate invariant subspaces have no invariant complement.
   (b) The factor is completely reducible, i.e. degenerate invariant subspaces have an invariant complement. This case occurs only for an indecomposable manifold of neutral signature (p,p)

Algebraically the situation of point 2. is analyzed in [Bou00]. About the corresponding indecomposable manifolds very few is known: in case of solvable symmetric spaces [CP80] also [Neu02], for index $(2, m-2)$ [Ike99] or for index (p,p) [BI97]. With respect to the question of parallel spinors there are investigations of I. Kath [Kat00] in neutral signature.

Now we turn to the Lorentzian manifolds, i.e we arrive at the topic of this thesis. We will try to answer the questions

(Q4) Which Lie groups can occur as holonomy group of a simply connected indecomposable, non-irreducible Lorentzian manifold?

(Q5) Which of these have trivial subrepresentations of the spinor representation?

In the remaining part of this introduction we will describe the content of this thesis and its results. Previous results of other authors are described in the corresponding parts of the thesis and cited in this context now.

In **chapter 1** above all we will recall the notions of bundles, connections, holonomy as well as the definition of the spin representations and the spin bundle. We will collect facts about semi-Riemannian holonomy, in particular the short Berger list. One section recalls the concept of local and infinitesimal holonomy as it was introduced by A. Nijenhuis [Nij53].
Furthermore we will prove that the product decomposition of the manifolds corresponds to a product decomposition of the spin bundles. In particular a product of semi-Riemannian manifolds admits a parallel spinor if and only if each of the factors admits a parallel spinor (corollary 1.22).
The remainder of this chapter applies this to Lorentzian manifolds after showing that a parallel spinor on a Lorentzian manifold defines a non-trivial parallel vector field on it. The fact that this vector field is causal (lemma 1.23) leaves us with two cases: it is timelike or lightlike.
In the timelike case the Lorentzian manifold splits due to the Wu-theorem in one flat (including the timelike invariant vector) factor and irreducible Riemannian factors. We can apply the theorem of Wang and get all such holonomy groups (theorem 1.24).
If the vector field is lightlike it defines a invariant degenerate subspace under holonomy representation, i.e. Wu-decomposition leads to irreducible or flat Riemannian factors with parallel spinors and one Lorentzian factor which is flat or proper indecomposable. Thus we have to deal with indecomposable, non-irreducible Lorentzian manifolds and its holonomy, the topic of chapter two.
A remarkable fact in this context is the one that a parallel spinor on a Lorentzian manifold implies that the holonomy of the manifold cannot be irreducible. This corresponds to the result that the only irreducible Lorentzian holonomy group is the full $SO(1, m-1)$. This is a consequence of the Berger list — the short and the long one — and was recently achieved in a direct and geometric way by J. Di Scala and C. Olmos [dSO01] and A. Zeghib and C. Boubel [Zeg03], [BZ03].

In the **chapter 2** we study indecomposable, non-irreducible Lorentzian manifolds. First of all we describe the basic algebraic properties of the holonomy. The holonomy group of an $n+2$ dimensional indecomposable, non-irreducible Lorentzian manifold is contained in the parabolic group $(\mathbb{R}^* \times SO(n)) \ltimes \mathbb{R}^n$ and defines projections on each of the three factors. If the holonomy is indecomposable the projection on $\mathbb{R}^n$ is surjective. The projection on $\mathbb{R}^*$ decides if the lightlike invariant subspace corresponds to a parallel lightlike vector field or only to a recurrent one. In both cases it defines a parallel isotropic one-dimensional distribution Ξ and its orthogonal complement $\Xi^\perp$, which is parallel too.
Then we prove that the $SO(n)$-projection of such a holonomy group is the holonomy group of a metrical vector bundle defined by the quotient of the parallel distributions

with restricted metric and equipped with a metrical connection defined by the Levi-Civita connection (corollary 2.6). This bundle is called *screen bundle*. A classification of such holonomy does not exist, but the geometric structures of the screen bundle correspond to algebraic properties of the $SO(n)$–projection of the holonomy.
In the following we cite the results of L. Berard-Bergery and A. Ikemakhen in [BI93]. On one hand they studied the algebraic structure of an indecomposable subalgebra of the parabolic algebra. Due to the relations between its projections they distinguished four different types, two of them are of the type $\mathfrak{g} \ltimes \mathbb{R}^n$ respectively $(\mathbb{R} \oplus \mathfrak{g}) \ltimes \mathbb{R}^n$ — we call them *uncoupled* — and for the remaining two hold relations between the center of the $\mathfrak{so}(n)$-projection and the $\mathbb{R}$ respectively the $\mathbb{R}^n$-part — the *coupled* types. (cited in theorem 2.10). There is one uncoupled and one coupled type with vanishing $\mathbb{R}$–component. We should remark that an analogous property is proved in the case of signature $(2, m-2)$ [Ike99].
On the other hand they proved a Borel-Lichnerowicz decomposition property for the $SO(n)$-projection of an indecomposable, non-irreducible Lorentzian holonomy group (cited in theorem 2.7). It will be very important in the following chapters. In particular it implies that the $SO(n)$-projection of the holonomy group is compact, although the whole holonomy group must not be compact. The latter occurs only for the types where the $SO(n)$ and the $\mathbb{R}^*$, resp. the $\mathbb{R}^n$-component are coupled. In contrary to the first result this one cannot be generalized to higher signature, since it uses the fact that the orthogonal complement of the invariant isotropic distribution has codimension one.
For sake of completeness we cite the classification of Lorentzian holonomy of four-dimensional space-times of [Sch60] and [Sha70].
Then we turn to the local shape of indecomposable, non-irreducible Lorentzian manifolds. It can be described with the help of the parallel distributions Ξ and $\Xi^\perp$ using the foliations of lines and lightlike hypersurface defined by them. Based one these, one finds coordinates — so-called Walker coordinates [Wal49] — with respect to these foliations. Furthermore we cite the theorem of C. Boubel [Bou00] which asserts the existence of coordinates adapted to the Borel-Lichnerowicz decomposition of the $SO(n)$-projection and gives conditions under which these coordinates are unique. With the help of these coordinates he gave a method to produce metrics of coupled type with the help of metrics of uncoupled type. With the results of [Bou00] the question (Q4) can be reduced to the question

(Q6) Which Lie groups can occur as $SO(n)$–component of the holonomy group of a simply connected indecomposable, non-irreducible Lorentzian manifold?

The Walker coordinates define families of Riemannian submanifolds of codimension 2. This implies, and is proved in [Ike96], that the holonomy groups of all these Riemannian submanifolds are contained in the $SO(n)$-projection of the Lorentz holonomy. In the remainder of this chapter we assume that these Riemannian manifolds have equivalent holonomy. In terms of the local connection form we describe the conditions under which the $SO(n)$-projection is equal to the holonomy of these Riemannian manifolds. Using this description we give sufficient conditions on the ingredients of the Walker-coordinate form of the metric such that the $SO(n)$-projection of the Lorentzian holonomy is a Riemannian holonomy. In corollary 2.37 we give the following construction method of indecomposable Lorentzian manifolds with holonomy $(\mathbb{R}^* \times$Riemannian holonomy$) \ltimes \mathbb{R}^n$ respectively (Riemannian holonomy) $\ltimes \mathbb{R}^n$:

2.37 Construction method: *Let (N, g) be a Riemannian manifold with holonomy G, γ a function of $\mathbb{R}$, $f \in C^\infty(N \times \mathbb{R}^2)$ a function, depending on every variable and ϕ_z a family of 1-forms on N, satisfying the property $d\phi_z \in \mathcal{H}(N, g) \times_G \mathfrak{g}$ where $\mathcal{H}(N, g)$ denotes the holonomy bundle of (N, g). Then the Lorentzian manifold*

$$\left(M = N \times \mathbb{R}^2, h = 2dxdz + fdz^2 + \phi_z dz + e^{2\gamma} \cdot g\right)$$

is indecomposable, non-irreducible and has holonomy $(\mathbb{R}^* \times G) \ltimes \mathbb{R}^n$ or $G \ltimes \mathbb{R}^n$ if $\frac{\partial}{\partial x}(f) = 0$.

Note that these sufficient conditions are very strong, since they require that the Riemannian metric *does not* depend on the parameter z. We apply this method to the case where the Riemannian manifold is Kähler, Ricci-flat Kähler or hyper-Kähler.

In **chapter 3** we will return to the problem of parallel spinors on indecomposable Lorentzian manifolds. The parallel spinor defines a parallel lightlike vector field, thus the holonomy has no $\mathbb{R}^*$-component, i.e. two of the four types of indecomposable subalgebras of the parabolic algebra due to Berard-Bergery and Ikemakhen are excluded. According to the shape of the holonomy algebra we will draw the following consequences for an indecomposable Lorentzian manifold of dimension $n + 2$:

Corollary 3.4. *Lorentzian manifolds with abelian holonomy $\mathbb{R}^n$ admit $2^{[\frac{n}{2}]}$ linear independent parallel spinors which are pure.*

Corollary 3.2. *For both types of holonomy, the coupled and the uncoupled type, the dimension of the space of parallel spinors does only depend on the $SO(n)$-component of the holonomy group.*

Under the assumption that the manifold admits a parallel spinor we can show for the $SO(n)$-projection of the holonomy:

Corollary 3.6. *All Riemannian metrics defined by the Walker coordinates have to admit parallel spinors.*

Theorem 3.11. *If it contains $SU(n/2)$ then it is equal to $SU(n/2)$.*

Proposition 3.14. *It cannot be abelian. Its irreducibly acting components have zero center or the dimension of the irreducible representations is divisible by 4.*

Corollary 3.6 restricts the possible class of Riemannian metrics occurring in the Walker coordinates. The only irreducible factors which can occur are these of the Wang list. Theorem 3.11 implies that an indecomposable Lorentzian manifold with parallel spinor whose Walker coordinates are defined by a family of $SU(n)$-metrics has holonomy $SU(n) \ltimes \mathbb{R}^n$.
In this context we cite the results of R. Bryant in low dimensions. He proved in [Bry99b] that for $n \leq 9$ holds the following: The *maximal* subgroups of $SO(n) \ltimes \mathbb{R}^n$ which fix a spinor are of type $G \ltimes \mathbb{R}^n$ where G is a Riemannian holonomy group with parallel spinor, i.e. product of entries of the Wang list and the unity. Furthermore we cite his results of [Bry99c] where the conditions on a family of $Spin(7)$-metrics to have Lorentzian holonomy $(1 \times Spin(7)) \ltimes \mathbb{R}^9$ are studied. See also [Bry00] and [FO00].
In the following we rewrite the equation of a parallel spinor in local coordinates, defined by a family of metrics with equivalent holonomy. If any metric in this family admits parallel spinors and if it satisfies the sufficient conditions of theorem 2.36 to have holonomy (Riemannian holonomy) $\ltimes \mathbb{R}^n$ then the Lorentzian manifold carries a parallel spinor. Analyzing the parallelism equation we can weaken these sufficient conditions (proposition 3.16). This gives further examples of local forms of Lorentzian manifolds with parallel spinors, but with holonomy (Riemannian holonomy)$\ltimes \mathbb{R}^n$.
Finally we study the particular class of indecomposable Lorentzian manifolds with parallel spinors for which the holonomy is abelian, i.e equals to $\mathbb{R}^n$. An example of these are the indecomposable Lorentzian symmetric spaces with solvable transvection group. With the help of theorem 2.36 we prove:

Theorem 3.19. *A simply connected indecomposable Lorentzian manifold with lightlike parallel vector field has abelian holonomy if and only if it is a pp-manifold.*

pp-manifolds are Lorentzian manifolds with a lightlike parallel vector field and an additional curvature condition (see [Sch74] and definition 3.18). They are generalizations of indecomposable Lorentzian symmetric spaces with solvable transvection group and generalizations of plane waves. *pp*-manifolds occur in physics as Penrose limits and were used in supergravity theories [BFOHP02].
At the end of chapter 3 we draw consequences from the fact that the Ricci tensor is totally isotropic, which was a necessary condition for the existence of a parallel spinor.

In **chapter 4** we try to answer the remaining question (Q6). This chapter is independent on the previous ones. The only fact we will use is the Borel-Lichnerowicz decomposition property of the $SO(n)$–component of an indecomposable, non-irreducible

Lorentzian holonomy group due to Berard-Bergery and Ikemakhen. Beside this property until here there is no further algebraic criterion. We try to answer (Q6) by evaluating the algebraic restrictions posed on the representation by the first Bianchi identity. Thus we formulate a criterion, which is analogous to the Berger criterion (2), but restricted to the $\mathfrak{so}(n)$-projection. We consider for a Lie algebra $\mathfrak{g} \subset \mathfrak{so}(E,h)$ (h is an inner product) the $\mathfrak{g}$-module

$$\mathcal{B}_h(\mathfrak{g}) := \{Q \in E^* \otimes \mathfrak{g} \mid h(Q(x)y,z) + h(Q(y)z,x) + h(Q(z)x,y) = 0\} \qquad (3)$$

and define:

Definition 4.2. *We call* $\mathfrak{g}$ ***weak-Berger*** *if* $span\{Q(x) \mid x \in E, Q \in \mathcal{B}_h(\mathfrak{g})\} = \mathfrak{g}$.

Weak-Berger algebras are generalizations of orthogonal Berger algebras. We prove

Theorem 4.6. *The* $\mathfrak{so}(n)$*-projection of an indecomposable, non-irreducible Lorentzian manifold is a weak-Berger algebra.*

Since two algebras (with its representations) are weak-Berger if and only if their sum is weak-Berger we can apply this criterion to the irreducible components of decomposition of the $SO(n)$-projection of an indecomposable Lorentzian holonomy.
Hence we have have to solve the following classification problem:

(Q7) Classify all irreducible weak-Berger algebras (with respect to a positive definite scalar product)!

Since the complexification of a weak-Berger algebra is also weak-Berger we will work with complex representation theory. But we have to split the problem into two cases:

1. The representation is of *real type*, i.e. its complexification is irreducible,

2. The representation is of *non-real type*, i.e. the complexification is not irreducible.

For the representations of real type (sections 4.2 and 4.3) the Lie algebra has to be semisimple. This enables us to use the tools of root space and weight space decomposition. We translate the weak-Berger property into a property of roots and weights, referring to the notion of spanning triples from [Sch99]. We arrive at slightly different conditions as in the Berger case. Restricting to the case where the Lie algebra is simple we prove the following

Theorem 4.51. *Any weak-Berger which is simple and of real type is a Berger algebra, in particular a Riemannian holonomy algebra.*

The proof uses an analysis completely analogously to the one in [Sch99], but with the weak-Berger property and imposing the condition that the representation is orthogonal.

For weak Berger algebras of non-real type (section 4.4) we have to deal with the problem that the complexification does not act irreducibly. We have to restrict to its irreducible components and describe the consequences for the scalar product and the weak-Berger property at this transition. Doing this, we will use the fact that the representation of the $SO(n)$-component is orthogonal with respect to a *positive definite* scalar product. Hence Lie algebras with representations of non-real type are contained in $\mathfrak{u}(n)$. For these we will prove that $\mathcal{B}_h(\mathfrak{g})^{\mathbb{C}}$ is isomorphic to the first prolongation of the complexification of $\mathfrak{g}$. The first prolongation of a Lie algebra $\mathfrak{g} \subset \mathfrak{gl}(V)$ is defined as

$$\mathfrak{g}^{(1)} := \{Q \in V^* \otimes \mathfrak{g} \mid Q(u)v = Q(v)u\}. \tag{4}$$

Using the classification result of Kobayashi and Nagano [KN65] for first prolongations of complex Lie algebras, we achieve the classification in question for Lie algebras which are contained in $\mathfrak{u}(n)$:

Theorem 4.62. *Any real irreducible weak-Berger algebra contained in $\mathfrak{u}(n)$ is a Berger algebra, in particular it is a Riemannian holonomy algebra.*

This result together with the result in the real-type case gives the

Conclusion: *Let G be the $SO(n)$-projection of the holonomy group of an indecomposable, non-irreducible $n+2$-dimensional Lorentzian manifold. If G is contained in $U(n)$, or its irreducible components are simple, then it is a Riemannian holonomy group. More detailed: Any irreducible component of G which is contained in $U(d)$ or simple is a Riemannian holonomy algebra.*

Of course all these algebras appearing in the conclusion can be realized by the $\mathfrak{so}(n)$–projection of an indecomposable non–irreducible Lorentzian manifold due to the construction method 2.31. To construct manifolds with holonomy of coupled type is possible due to the result of [Bou00], but with our result only under the assumption that one irreducible factor equals to a Riemannian holonomy algebra with center, i.e. equal to $\mathfrak{u}(d)$ or the holonomy algebra of a Riemannian hermitian symmetric space. Regarding parallel spinors again, we get the following:

Corollary 4.65. *If an indecomposable, non-irreducible Lorentzian manifold carries a parallel spinor, then its reduced holonomy group is of uncoupled type $G \ltimes \mathbb{R}^n$.*

Actually these results leave open the question: Are there semisimple, non simple, irreducibly acting Lie algebras, not contained in $\mathfrak{u}(n/2)$, which are weak-Berger, but not Berger? We guess that this is not the case, because the methods for weak-Berger algebras of real type will work also in the proper semisimple case, but this is still work in progress. Up to dimension $n \leq 9$ this was proved very recently by A. Galaev [Gal03] for algebras not contained in $\mathfrak{u}(n/2)$. In his paper he studied the space of curvature endomorphisms for indecomposable subalgebras in $(\mathbb{R} \oplus \mathfrak{so}(n)) \ltimes \mathbb{R}^n$. Reducing everything to one uncoupled type he proved the other direction of our result: a subalgebra of one of these types is a Berger algebra, if its $\mathfrak{so}(n)$–projection is a weak-Berger algebra (in our terms).

We are aware that the proofs we will present in this section are a cumbersome case-by-case analysis using the methods of representation theory. It is very desirable to get a direct and more geometric proof of the proposition that every $SO(n)$–projection of an indecomposable, non-irreducible Lorentzian holonomy group is a Riemannian holonomy group, which includes the remaining semisimple case of course. A further interesting question — maybe connected to a geometric proof of the above fact — is to ask for the most general form of the metric having holonomy in $(\mathbb{R}^* \times SO(n)) \ltimes \mathbb{R}^n$ with given Riemannian holonomy as $SO(n)$–projection.

Chapter 1

Holonomy and parallel spinors

In this chapter we will recall the notions of bundles, connections, holonomy as well as the definition of the spin representations and the spin bundle. The relation between parallel sections in vector bundles and trivial subrepresentations of the holonomy representation is explained. We will collect facts about semi-Riemannian holonomy, the splitting theorem of de-Rham and Wu and the Berger list of non-symmetric semi-Riemannian manifolds. One section recalls the concept of local and infinitesimal holonomy introduced by A. Nijenhuis. Furthermore we will prove that a product of semi-Riemannian manifolds admits parallel spinors if and only if each of the factors admits parallel spinors. This will apply to the Lorentzian case where the parallel spinor induces a causal parallel vector field. The case where this vector field is lightlike remains as problematic case.

1.1 Holonomy - Definition and basic facts

1.1.1 Holonomy of a connections in a principal fibre bundle

Let $\pi : P \to M$ be a principal fibre bundle with structure group G and $\omega : TP \to \mathfrak{g} := LA(G)$ a connection in P.
The horizontal space is defined as $Th_pP := Ker\ \omega_p$. This enables us to assign to every curve γ in M its horizontal lift γ^* in P with $(\gamma^*)'(t) \in Th_{\gamma^*(t)}P$ and a fixed starting point. The parallel displacement $\mathfrak{p}^{\omega}_{\gamma}$ along a piecewise smooth curve γ in M from $P_{\gamma(0)}$ to $P_{\gamma(1)}$ assigns $\gamma^*(0)$ the endpoint of the horizontal lift $\gamma^*(1)$.
Since G acts freely and transitively on the fibres, one can define the holonomy group of ω in a point $p \in P$ with $\pi(p) = x \in M$:

$$Hol_p(P,\omega) := \left\{ A \in G \;\middle|\; \begin{array}{l} \text{there is a piecewise smooth curve } \gamma \text{ in } M \text{ with} \\ \gamma(0) = \gamma(1) = x \text{ and } R_A\left(\gamma^*_p(0)\right) = \gamma^*_p(1). \end{array} \right\} \subset G.$$

The restricted holonomy group $Hol^0_p(P,\omega)$ is defined by restricting the definition to homotopically trivial curves in M. This group is connected and lies in the connected component of the unity in G. Its Lie algebra is denoted by $\mathfrak{hol}_p(P,\omega) := LA\left(Hol^0_p(P,\omega)\right)$. The holonomy groups depend in the following way on the point $p \in P$: for two points p and $q = R_A(p)$ in the same fibre over $x \in M$, with $A \in G$, the holonomy groups are conjugated in G, i.e. $Hol_q(P,\omega) = A^{-1}Hol_p(P,\omega)A$. If there is a horizontal curve in P from p to q then the holonomy groups are the same.
For the curvature Ω of ω and the group action holds the following

$$R_A^*\Omega = Ad\left(A^{-1}\right) \circ \Omega. \tag{1.1}$$

In order to localize the connection form and the curvature one considers a local section $s : U \to P$. Then the local connection and curvature form are given by $\omega^s := s^*\omega = \omega \circ ds$ and $\Omega^s := s^*\Omega$.

Reduction theorem and holonomy theorem Let P be a principal G-fibre bundle and $\lambda : H \to G$ be a Lie group homomorphism. A principal H-fibre bundle Q and a smooth mapping $f : Q \to P$ are called **λ-reduction** of P if the following diagram commutes

$$\begin{array}{ccccc} H \times Q & \longrightarrow & Q & & \\ & & & \searrow & \\ \lambda\times f\downarrow & \circlearrowleft & f\downarrow & \circlearrowleft & M. \\ & & & \nearrow & \\ G \times P & \longrightarrow & P & & \end{array}$$

If λ is the injection of a Lie subgroup H one calls Q a reduction of P. Now we recall the notion of a reduction of a connection. Let P a principal G-fibre bundle and Q a $(\lambda : H \to G)$-reduction of P. If now θ is a connection in Q then there exists a unique connection ω of P with the property $df_q(Th^\theta_q Q) \subset Th^\omega_{f(q)}P$.

Furthermore the right hand side diagram commutes, i.e. $f^*\omega = \lambda_* \circ \theta$ and for the curvatures Θ and Ω holds $f^*\Omega = \lambda_* \circ \Theta$. Then ω is called **λ-extension of** θ and θ is called **λ-reduction of** ω.

$$\begin{array}{ccc} TQ & \xrightarrow{df} & TP \\ \theta\downarrow & & \downarrow\omega \\ \mathfrak{h} & \xrightarrow[\lambda_*]{} & \mathfrak{g} \end{array}$$

An extension exists for every λ but not a reduction. A criterion gives the following

1.1 Proposition. *Let $H \subset G$ be a closed Lie subgroup such that the Lie algebras are a reductive pair, i.e. there is a subspace $\mathfrak{m} \subset \mathfrak{g}$ such that $\mathfrak{g} = \mathfrak{h} \oplus \mathfrak{m}$ and $Ad(H)\mathfrak{m} \subset \mathfrak{m}$. If furthermore P is a principal G-fibre bundle with connection ω and $f : Q \to P$ is a $H \subset G$-reduction, then*

$$\theta := pr_{\mathfrak{h}} \circ \omega_{|TQ} : TQ \longrightarrow \mathfrak{h}$$

is a connection on Q. If $\omega_{|TQ} : TQ \to \mathfrak{h}$ then θ is a reduction of ω.

The reduction theorem now asserts that there is a smallest reduction of a connection on the holonomy bundle. The **holonomy bundle** of a principal fibre bundle P with connection ω in a point $p \in P$ is defined as follows

$$\mathcal{H}ol^{\omega}(p) := \{q \in P | \text{ There is a } \omega\text{-horizontal curve from } p \text{ to } q\}$$

Then it holds:

1.2 Theorem (Reduction theorem). *Let P be a principal fibre bundle over a connected manifold with structure group G and connection ω. Let $p \in P$. Then the holonomy bundle $\mathcal{H}ol^{\omega}(p)$ is a principal fibre bundle with structure group $Hol_p(\omega)$ and a reduction of P and ω.*

This theorem is used to prove the second main result of holonomy theory, the Ambrose-Singer-holonomy-theorem.

1.3 Theorem (Holonomy theorem). *[AS53] Let M be connected and P a principal fibre bundle over M with structure group G and connection ω. Then*

$$\mathfrak{hol}_p(P,\omega) = \left\{ \Omega_q(X,Y) \;\middle|\; \begin{array}{c} \text{there is a } \omega\text{-horizontal curve from } p \text{ to } q \\ X, Y \in Ker\ \omega_q \subset T_qP \end{array} \right\} \subset \mathfrak{g}. \qquad (1.2)$$

In the terms of the right hand side of (1.2) only horizontal vectors are needed. This enables us to calculate these terms via formula (1.1) with the help of the local curvature form:

$$\Omega_q(X,Y) = Ad\left(g^{-1}\right)\Omega_{s(x)}\left(dR_{g^{-1}}(X), dR_{g^{-1}}(Y)\right) = Ad\left(g^{-1}\right)\Omega^s_x(\hat{X},\hat{Y}), \qquad (1.3)$$

where $q = R_g(s(x))$ on a horizontal curve from p and $dR_g\left(ds(\hat{X})\right) = X$, $dR_g\left(ds(\hat{Y})\right) = Y$ both horizontal.

1.1.2 Holonomy and parallel sections in vector bundles

Let (V,ρ) a representation of G. Now one consider the vector bundle associated to P and G defined by $E := (P \times V)/G$. Fixing a $p \in P_x$ gives an isomorphism of G acting on V onto a subgroup of $Gl(E_x)$:

$$\begin{array}{rccl} \Psi_p & : & G \subset Gl(V) & \hookrightarrow & Gl(E_x) \\ & & A & \mapsto & ([p,v] \mapsto [p,\rho(A)v] \in E_x), \end{array}$$

i.e. a representation of G on E_x. For a different $q = R_{B^{-1}}(p) \in P_x$ it is

$$\Psi_p(A)[p,v] = [p,\rho(A)v] = [R_B(q),\rho(A)v] = [q,\rho(B^{-1}A)v],$$

i.e. $\Psi_p(A)[q,w] = [q, \rho(B^{-1}AB)v]$. Hence the representations Ψ_p and Ψ_q on E_x are conjugate to each other with respect to G in the sense

$$\Psi_p(A) = \Psi_q(B^{-1}AB). \tag{1.4}$$

ω defines a covariant derivative by the formula $\nabla^\omega_X e = [s(x), \rho_*(\omega^s(X))(v) + dv(X)]$, where $s : U \to P$ is a local section in P and $e(x) := [s(x), v(x)]$ a section in the vector bundle. The corresponding parallel displacement of ∇^ω along a curve γ from x to y in M, denoted by $\mathcal{P}^{\nabla^\omega}_\gamma : E_x \to E_y$, then is defined as $\mathcal{P}^{\nabla^\omega}_\gamma(e) = \mathcal{P}^{\nabla^\omega}_\gamma([p,v]) := [\mathfrak{p}^\omega_\gamma(p), v]$. It satisfies for a loop γ around x

$$\mathcal{P}^{\nabla^\omega}_\gamma([p,v]) = [\mathfrak{p}^\omega_\gamma(p), v] = [R_A(p), v] = [p, \rho(A^{-1})(v)] \tag{1.5}$$

for all $e \in E_x$. Now one can define the holonomy group in a point $x \in M$ with respect to this connection acting on the fibre E_x:

$$Hol_x(E, \nabla^\omega) := \{\mathcal{P}^{\nabla^\omega}_\gamma | \gamma : [0,1] \to M,\ \gamma(0) = \gamma(1) = x\} \subset Gl(E_x).$$

The restricted holonomy group $Hol^0_x(E, \nabla^\omega)$ is defined as a subgroup of $Hol_x(E, \nabla^\omega)$ by restricting the possible curves to those which are homotopic to the constant curve. Because of formula (1.5) the identification of G with a subgroup of $Gl(E_x)$ gives an identification between the holonomy groups and their representations

$$(Hol_p(P,\omega), V) \overset{\Psi_p}{\simeq} (Hol_x(E, \nabla^\omega), E_x) \quad \text{for } p \in P_x.$$

In this sense one refers to the holonomy representation. Thus with (1.4) the holonomy representations of the holonomy groups of two different points in M can be understood as representations which are conjugated with respect to an element of G.
Now we can identify the parallel sections in E with invariant vectors under the holonomy representation:

$$\begin{array}{rcl} V_{Hol_q(\omega)} := \{v \in V \,|\rho(Hol_q(\omega))(v) = v\} & \simeq & \{e \in \Gamma(E) | \nabla^\omega e = 0\} \\ v & \mapsto & [\mathfrak{p}^\omega_\gamma(p), v] \end{array} \tag{1.6}$$

Or in case that M is simply connected one has on the Lie-algebra level

$$V_{\mathfrak{hol}_q(\omega)} := \{v \in V \,|\rho_*(\mathfrak{hol}_q(\omega))(v) = 0\} \simeq \{e \in \Gamma(E) | \nabla^\omega e = 0\}. \tag{1.7}$$

1.1.3 Local and infinitesimal holonomy group

In this section we will recall some results of Albert Nijenhuis about the local and infinitesimal holonomy group and its relations to the reduced one. Nijenhuis proved the Ambrose-Singer holonomy theorem before Ambrose and Singer in his book [Nij52] but only for affine and analytic manifolds. As a consequence of this lack he introduced in [Nij53] the notions of local and infinitesimal holonomy group. We cite these results here for connections in arbitrary principle fibre bundles following [KN63], pp 94 .

The local holonomy group We start with a connected manifold M, a principal fibre bundle P and a connection ω. If U is an open set we can restrict the bundle P to U, denoted by $P_{|U}$. Also the connection can be restricted to $P_{|U}$.

1.4 Definition. $Hol^*_p(\omega) := \bigcap_{U \ni \pi(p)} Hol^0_p(P_{|U}, \omega)$ is called **local holonomy group.**

For $Hol^*_p(\omega)$ the following properties are valid.

1.5 Proposition. *1. $Hol^*_p(\omega)$ is a connected Lie group. Furthermore it is a Lie subgroup of $Hol^0_p(P, \omega)$.*

2. There is a neighborhood U of $\pi(p)$ such that $Hol^*_p(\omega) = Hol^0_p(P_{|U}, \omega)$.

3. In a fibre holds $Hol^*_{A \cdot p}(\omega) = A^{-1}\left(Hol^*_p(\omega)\right) A$ for $A \in G$.

Now Nijenhuis proved the following

1.6 Theorem. *[Nij53] Let M be a connected manifold and P a principal fibre bundle with connection ω. Then it holds for the local and the reduced holonomy in a point $p \in P$*

1. *$Hol^0_p(P, \omega)$ is generated by elements of $Hol^*_q(\omega)$ with $q \in \mathcal{H}ol^\omega(p)$.*
2. *$\mathfrak{hol}^0_p(P, \omega)$ is spanned by elements of $\mathfrak{hol}^*_q(\omega)$ for $q \in \mathcal{H}ol^\omega(p)$.*
3. *If the dimension of $Hol^*_q(\omega)$ is constant over M, then $Hol^0_p(P, \omega) = Hol^*_p(\omega)$.*

The infinitesimal holonomy group and algebra To define the infinitesimal holonomy algebra we consider the curvature form Ω of a connection ω in a principal fibre bundle P with structure group G. If $\mathfrak{g}$ is the Lie algebra of G we consider the sequence of subspaces of $\mathfrak{g}$:

$$\begin{aligned}
\mathfrak{h}^{(0)}_p &:= span\left(\Omega(X,Y) | X, Y \text{ horizontal vector fields of } P \text{ around } p\right) \\
&\vdots \\
\mathfrak{h}^{(k)}_p &:= span\left(\mathfrak{h}^{(k-1)}_p \cup \left\{ \begin{array}{l} Z_k \cdots Z_1(\Omega_p(X,Y)) | \\ Z_1, \ldots Z_k, X, Y \text{ horizontal vectors fields of } P \text{ around } p \end{array} \right\}\right)
\end{aligned}$$

Then it holds the relation

$$\left[\mathfrak{h}^{(k)}_p, \mathfrak{h}^{(l)}_p\right] \subset \mathfrak{h}^{(k+l+2)}_p,$$

which turns the space

$$\mathfrak{hol}'_p(\omega) := \bigcup_{k=0,1,\ldots} \mathfrak{h}^{(k)}_p$$

into a Lie algebra. Furthermore $\mathfrak{hol}'_p(\omega)$ is a Lie subalgebra of $\mathfrak{hol}^*_p(\omega)$.

1.7 Definition. The connected Lie subgroup $Hol'_p(\omega)$ of $Hol^*_p(\omega)$ with Lie algebra $\mathfrak{hol}'_p(\omega)$ is called **infinitesimal holonomy group.**

1.8 Proposition. *1. $Hol'_p(\omega)$ is a connected Lie subgroup of $Hol^*_p(\omega)$.*

2. In a fibre holds $Hol'_{A\cdot p}(\omega) = Ad(A^{-1})\left(Hol'_p(\omega)\right)$ for $A \in G$.

Thus we have the following inclusions of connected Lie subgroups

$$Hol'_p(\omega) \subset Hol^*_p(\omega) \subset Hol^0_p(P,\omega) \subset Hol_p(P,\omega).$$

1.9 Theorem. *[Nij53] Let $Hol'_p(\omega)$ be the infinitesimal holonomy group of a connection ω in a principal fibre bundle P over a connected manifold M. Then it holds*

1. *If the dimension of $Hol'_p(\omega)$ is constant in a neighborhood of p in P then $Hol'_p(\omega) = Hol^*_p(\omega)$. If it is constant on P then $Hol'_p(\omega) = Hol^0_p(P,\omega)$.*
2. *If P and ω are real analytic, then $Hol'_p(\omega) = Hol^*_p(\omega) = Hol^0_p(P,\omega)$ for every $p \in P$.*

1.1.4 Holonomy of semi-Riemannian manifolds

Let (M,h) be a semi-Riemannian manifold of dimension $m = r+s$, index r and ∇ the Levi-Civita-connection.
$\mathcal{O}(M,h)$ denotes the bundle of orthonormal frames over M with structure group $O(r,s)$. Then one has as in section 1.1.2: $\mathcal{O}(M,h) \times_{O(r,s)} \mathbb{R}^m \simeq TM$.
The Levi-Civita connection ∇ defines a connection ω in $\mathcal{O}(M,h)$. Its local connection form is given by

$$\omega^s = \sum_{1\leq i<j\leq n} h\left(\nabla s_i, s_j\right) D_{ij},$$

where $s = (s_1, \ldots s_m)$ is a section in P, that is a local orthonormal frame-field and D_{ij} the standard basis in $\mathfrak{o}(r,s)$. Then is $\nabla = \nabla^\omega$. One defines the holonomy representations

$$\begin{array}{lcccc} (Hol_x(M,h), T_xM) & := & (Hol_x(TM,\nabla), T_xM) & \simeq & (Hol_p(\mathcal{O}(M,h),\omega), \mathbb{R}^m) \\ \left(Hol^0_x(M,h), T_xM\right) & := & \left(Hol^0_x(TM,\nabla), T_xM\right) & \simeq & \left(Hol^0_p(\mathcal{O}(M,h),\omega), \mathbb{R}^m\right) \end{array}.$$

The relation between parallel vector fields and invariant subspaces of the holonomy representation is given by the following theorem (see for example [Bes87])

1.10 Theorem. *Let (M,h) be a semi-Riemannian manifold. Then it holds the following*

1. *For $1 \leq k < m$ these propositions are equivalent:*

(i) *There is a k-dimensional distribution, which is invariant under parallel displacement.*

(ii) *The holonomy representation leaves invariant a k-dimensional subspace.*

This distribution has to be involutive.

2. *There is a parallel vector field on M if and only if the holonomy representation has a trivial subrepresentation, that means it fixes a vector.*

The properties of the representation are used to characterize the manifold.

1.11 Definition. A semi-Riemannian manifold $(M^{r,s}, h)$ is called

1. **(strictly) irreducible** if the (reduced) holonomy representation has no invariant subspace,
2. **(strictly) indecomposable** if the (reduced) holonomy representation has no invariant subspace on which h is non-degenerate,
3. **reducible/decomposable** if it is not irreducible/indecomposable.

Irreducibility entails indecomposability and decomposability reducibility. For Riemannian manifolds both notions are the same. The tangential space of a semi-Riemannian manifold can be decomposed in a orthogonal sum of invariant subspaces which can be (in case of nontrivial signature of h) reducible with a degenerate, invariant subspace. For a given semi-Riemannian manifolds $(M_1^{r,s}, h_1)$ and $(M_2^{p,q}, h_2)$ the product manifold $(M_1^{r,s} \times M_2^{p,q}, h_1 \oplus h_2)$ has the tangential space $T_{(x_1,x_2)}(M_1 \times M_2) = T_{x_1}M_1 \oplus T_{x_2}M_2$. Now the holonomy representation of the product is the product of the holonomy representations (see again [Bes87]):

$$\begin{aligned} Hol_{(x_1,x_2)}(M_1 \times M_2, h_1 \oplus h_2) &= Hol_{x_1}(M_1, h_1) \times Hol_{x_2}(M_2, h_2) \\ &= (Hol_{x_1}(M_1, h_1) \times 1) \oplus (1 \times Hol_{x_2}(M_2, h_2)) . \end{aligned}$$

The de-Rham decomposition theorem asserts under which conditions the converse is true.

1.12 Theorem. *[dR52] [Wu64] Every simply connected, complete semi-Riemannian manifold is isometric to a product of simply connected, complete manifolds, which are flat or indecomposable.*

This means, locally it is sufficient to know the indecomposable manifolds. For the subclass of irreducible ones there is the classification of Berger and Simons.

1.13 Theorem (Non-symmetric Berger list). *[Ber55], [Sim62], [Ale68], [BG72] and [Bry87] Let $(M^{r,s}, h)$ be a simply connected, irreducible semi-Riemannian manifold of dimension $m = r+s$ and index r, which is not locally-symmetric. Then the holonomy representation on $\mathbb{R}^m$ is one of the following (modulo conjugation in $O(r,s)$):*

$$\begin{array}{llll}
m = r+s \geq 2 & : & SO(r,s) & \\
m = 2p+2q \geq 4 & : & U(p,q) \text{ or } SU(p,q) & \subset SO(2p,2q) \\
m = 4p+4q \geq 8 & : & Sp(p,q) \text{ or } Sp(p,q)\cdot Sp(1) & \subset SO(4p,4q) \\
m = r+r \geq 4 & : & SO(r,\mathbb{C}) & \subset SO(r,r) \\
m = 2p+2p \geq 8 & & Sp(p,\mathbb{R})\cdot Sl(2,\mathbb{R}) & \subset SO(2p,2p) \\
m = 4p+4p \geq 16 & & Sp(p,\mathbb{C})\cdot Sl(2,\mathbb{C}) & \subset SO(4p,4p) \\
m = 7 = 0+7 & : & G_2 & \subset SO(7) \\
m = 7 = 4+3 & : & G^*_{2(2)} & \subset SO(4,3) \\
m = 14 = 7+7 & : & G_2^{\mathbb{C}} & \subset SO(7,7) \\
m = 8 = 0+8 & : & Spin(7) & \subset SO(8) \\
m = 8 = 4+4 & : & Spin(4,3) & \subset SO(4,4) \\
m = 16 = 8+8 & : & Spin(7)^{\mathbb{C}} & \subset SO(8,8)
\end{array}$$

In case of symmetric spaces there is a classification of Berger [Ber57], which gives the following corollary for Lorentzian manifolds.

1.14 Corollary. *A simply connected, complete and irreducible Lorentzian manifold has the the full $SO_0(1, m-1)$ as holonomy group.*

This corollary was proved in a geometric way independently on the Berger list by J. Di Scala and C. Olmos [dSO01]. Recently it was proved by A. Zeghib [BZ03], who used dynamical methods.

1.2 Parallel Spinors on semi-Riemannian manifolds

1.2.1 Spinor representations and the spinor bundle

Clifford algebra and spin groups Now we want to define the spinor bundle of an $m = (r+s)$–dimensional semi-Riemannian manifold. (See [LM89], [Bau81] also [Bau94] for details.) Let $Cl_{r,s}$ be the Clifford algebra of $(\mathbb{R}^m, \langle .,. \rangle_{r,s})$ where $\langle .,. \rangle_{r,s}$ is the standard scalar product of index r on $\mathbb{R}^m$. We define the following groups in $Cl_{r,s}$ ($\langle X, X\rangle_{r,s} =: \|X\|_{r,s}$):

$$\begin{aligned}
Spin(r,s) &:= \{X_1 \cdot \ldots \cdot X_{2k} \mid \|X_i\|_{r,s} = \pm 1,\ k \geq 0\} \\
Spin_0(r,s) &= \{X_1 \cdot \ldots \cdot X_{2k} \cdot Y_1 \cdot \ldots \cdot Y_{2l} \mid \|X_i\|_{r,s} = 1, \|Y_i\|_{r,s} = -1,\ k,l \geq 0\}
\end{aligned}$$

$$K = \left\{ X_1 \cdot \ldots \cdot X_{2k} \cdot Y_1 \cdot \ldots \cdot Y_{2l} \;\middle|\; \begin{array}{l} \|X_i\|_{r,s} = 1, \|Y_i\|_{r,s} = -1, \quad k,l \geq 0 \\ X_i \in span(e_1, \ldots, e_r),\ Y_j \in span(e_{r+1}, \ldots, e_m) \end{array} \right\},$$

where $(e_1, \ldots e_m)$ is a basis with $\langle e_i, e_j \rangle = \delta_{ij}\kappa_i$, $\kappa_1 = \ldots = \kappa_r = -1$ and $\kappa_{r+1} = \ldots = \kappa_m = 1$. $Spin_0(r,s)$ is the connected component of the unit and K its maximal compact subgroup.
Now let $\lambda : Spin(r,s) \to SO(r,s)$ be the twofold covering of $SO(r,s)[SO_0(r,s)]$ by $Spin(r,s)[Spin_0(r,s)]$. Its differential λ_* is a Lie algebra isomorphism between $\mathfrak{spin}(r,s) := LA(Spin(r,s)) \subset Cl_{r,s}$ and $\mathfrak{so}(r,s)$. Then

$$\mathfrak{spin}(r,s) = span\{e_i \cdot e_j | 1 \leq i < j \leq m\}$$

and

$$\lambda_*(e_i \cdot e_j) = D_{ij} \quad \text{for matrices } D_{ij} = -\kappa_j E_{ij} + \kappa_i E_{ji} \tag{1.8}$$

which form a basis in $\mathfrak{so}(r,s)$, where E_{ij} is the standard basis of $\mathfrak{gl}(m, \mathbb{R})$ with the (i,j)–th entry 1 and all other zero.

A representation of the Clifford algebra and the spin group Now we will give an isomorphism between the complexification of the Clifford algebra $\mathbb{C}l_{(r,s)} := Cl_{r,s} \otimes \mathbb{C}$ and endomorphism algebras of complex vector spaces which yields to complex representations of the spin group. First we consider the $\mathbb{C}^2$ with a basis $\left\{ u(\varepsilon) := \frac{1}{\sqrt{2}} \binom{1}{-\varepsilon i} | \varepsilon = \pm 1 \right\}$ and the isomorphisms of $\mathbb{C}^2$:

$$E := Id \ , \ T := \begin{pmatrix} 0 & -i \\ i & 0 \end{pmatrix} \ , \ U := \begin{pmatrix} i & 0 \\ 0 & -i \end{pmatrix} \ , \ V := \begin{pmatrix} 0 & i \\ i & 0 \end{pmatrix}.$$

Then we have $T^2 = -V^2 = -U^2 = E$, $UT = -iV$, $VT = iU$, $UV = -iT$ and $Tu(\varepsilon) = -\varepsilon u(\varepsilon)$, $Uu(\varepsilon) = iu(-\varepsilon)$, $Vu(\varepsilon) = \varepsilon u(-\varepsilon)$. We define the isomorphisms as follows:

1. In case m is even $\Phi_{(r,s)} : \mathbb{C}l_{(r,s)} \to \mathbb{C}\left(2^{\frac{m}{2}}\right)$ is defined by

$$\begin{aligned} \Phi_{(r,s)}(e_{2k-1}) &:= \tau_{2k-1} E \otimes \ldots \otimes E \otimes U \otimes \underbrace{T \otimes \ldots \otimes T}_{(k-1)\text{-times}} \\ \Phi_{(r,s)}(e_{2k}) &:= \tau_{2k} E \otimes \ldots \otimes E \otimes V \otimes \underbrace{T \otimes \ldots \otimes T}_{(k-1)\text{-times}} \end{aligned} \tag{1.9}$$

   with $\tau_1 = \ldots = \tau_r = i$ and $\tau_{r+1} = \ldots = \tau_m = 1$ and $k = 1 \ldots \frac{m}{2}$.

2. In case m is odd $\Phi_{(r,s)} : \mathbb{C}l_{(r,s)} \to \mathbb{C}\left(2^{\frac{m-1}{2}}\right) \oplus \mathbb{C}\left(2^{\frac{m-1}{2}}\right)$ is defined by

$$\begin{aligned} \Phi_{(r,s)}(e_k) &= \left(\Phi_{(r,s-1)}(e_k), \Phi_{(r,s-1)}(e_k)\right) \ , \ k = 1 \ldots m-1 \\ \Phi_{(r,s)}(e_m) &= (iT \otimes \ldots \otimes T, -iT \otimes \ldots \otimes T). \end{aligned}$$

This yields representations of the spin group and algebra in case m even by restriction and in case m odd by restriction and projection onto the first component. The representation space $\Delta_{r,s} = \mathbb{C}^{2^{[\frac{m}{2}]}}$ is called **spinor module.** We write $A \cdot v := \Phi_{(r,s)}(A)(v)$ for $A \in \mathcal{C}l_{r,s}$ and $v \in \Delta_{r,s}$. A useful basis in $\Delta_{r,s}$ is the following:

$$(u\,(\varepsilon_k, \dots, \varepsilon_1) := u(\varepsilon_k) \otimes \dots \otimes u(\varepsilon_1) | \varepsilon_i = \pm 1)\,.$$

In case m is even the $Spin(r,s)$–representation space $\Delta_{r,s}$ splits into two irreducible subspaces $\Delta^{\pm}_{r,s} := span(u\,(\varepsilon_k, \dots, \varepsilon_1)\,|\varepsilon_k \cdot \dots \cdot \varepsilon_1 = \pm 1)$.

Hermitian inner products on the spinor module On $\mathbb{C}^{2^{[\frac{m}{2}]}}$ we consider first the standard hermitian product $(v,w) := \sum_{k=1}^{2^{[\frac{m}{2}]}} v_k \cdot \overline{w}_k$, which is positive definite and invariant under the action of the maximal compact subgroup K of $Spin_0(r,s)$. If $\theta : \mathbb{R}^m \mapsto \mathbb{R}^m$ denotes the reflection on the spacelike subspace $span\ (e_{r+1}, \dots, e_m)$ then for $(.,.)$ holds the relation

$$(X \cdot v, w) + (v, \theta(X) \cdot w) = 0\ , \qquad \text{for } v, w \in \Delta_{r,s}\ ,\ X \in \mathbb{R}^m.$$

The basis given by $u\,(\varepsilon_k, \dots, \varepsilon_1)$ with $k := [\frac{m}{2}]$ is an orthonormal basis for $(.,.)$.
To define the second hermitian product we consider the element of the Clifford algebra

$$a := i^{\frac{r(r-1)}{2}} e_1 \cdot \dots \cdot e_r.$$

a satisfies the relation $(a \cdot v, w) = (v, a \cdot w)$. This leads to the definition of the second hermitian product

$$\langle u, v\rangle := (a \cdot u, v),$$

since $\langle u, v\rangle = (a \cdot u, v) = (u, a \cdot v) = \overline{(a \cdot v, u)} = \overline{\langle v, u\rangle}$. In case $r > 0$ it is indefinite of index $2^{[\frac{m}{2}]-1}$ and in case $r = 0$ equal to the positive definite $(.,.)$. It is invariant under $Spin_0(r,s)$ and satisfies the relation

$$\langle X \cdot v, w\rangle + (-1)^r \langle v, X \cdot w\rangle = 0,$$

which implies that $i^{r+1}\langle X \cdot v, v\rangle$ is a real number.

The spinor bundle Consider a semi-Riemannian manifold which is assumed to be spin. Let $(\widetilde{\mathcal{O}}, f)$ be the λ-reduction for the orthonormal frame bundle $\mathcal{O}(M,h)$. Then we have $TM = \mathcal{O}(M,h) \times_{SO_0(r,s)} \mathbb{R}^m = \widetilde{\mathcal{O}} \times_{Spin_0(r,s)} \mathbb{R}^m$. The vector bundle

$$S := \widetilde{\mathcal{O}} \times_{Spin_0(r,s)} \Delta_{r,s}$$

is called **spinor bundle.** The Clifford multiplication is defined as follows

$$\begin{array}{rcl} TM \otimes S = \left(\widetilde{\mathcal{O}} \times_{Spin_0(r,s)} \mathbb{R}^m\right) \otimes \left(\widetilde{\mathcal{O}} \times_{Spin_0(r,s)} \Delta_{r,s}\right) & \to & S \\ X \otimes \varphi = [q,x] \otimes [q,v] & \mapsto & [q, X \cdot v] =: X \cdot \varphi \end{array}$$

Since the indefinite scalar product $\langle .,. \rangle$ is invariant under $Spin_0(r,s)$ it defines a scalar product on S. But one can also define the positive definite scalar product invariantly by fixing a timelike (negative definite) subbundle $\xi \subset TM$. Then there is a $K \subset Spin_0(r,s)$–reduction of $\widetilde{\mathcal{O}}$ to a K–principal bundle $\widetilde{\mathcal{O}}_\xi$ for which

$$S = \widetilde{\mathcal{O}} \times_{Spin_0(r,s)} \Delta_{r,s} = \widetilde{\mathcal{O}}_\xi \times_K \Delta_{r,s}.$$

Thus the K invariance of $(.,.)$ suffices to define positive definite scalar product $(.,.)_\xi$ on S.

The covariant derivative defined by the Levi-Civita connection is given by

$$\nabla^S_X \varphi = X(\varphi) + \frac{1}{2}\sum_{i<j} \kappa_i\ \kappa_j\ g(\nabla^{LC}_{s_i} X, s_j) s_i \cdot s_j \cdot \varphi. \tag{1.10}$$

∇^S is metrical with respect to $\langle .,. \rangle$ and distributive in the following way

$$\nabla^S (X \cdot \varphi) = (\nabla X) \cdot \varphi + X \cdot \nabla^S \varphi.$$

For its holonomy group we have the important relation to those of ∇

$$\lambda \left(Hol_x(S, \nabla^S \right) = Hol_x(M,h).$$

Hence λ_* identifies the Lie algebras $\mathfrak{hol}_x(S,\nabla^S) \overset{\lambda_*}{\simeq} \mathfrak{hol}_x(M,h)$.

1.2.2 First properties of parallel spinors

Let (M,h) be a semi-Riemannian spin manifold of dimension m. We define

1.15 Definition. A non-trivial spinor field $\varphi \in \Gamma(S)$ is called **parallel** if $\nabla^S \varphi = 0$.

If we set $\tilde{H} := \lambda^{-1}(Hol_x(M,h))$ and $\tilde{\mathfrak{h}} := \lambda_*^{-1}(\mathfrak{h})$, then we have from the previous section

$$\{\text{parallel spinors}\} \simeq V_{\tilde{H}} := \left\{ v \in \Delta_{r,s} \middle| (\Phi_{(r,s)})(\tilde{H})(v) = v \right\}$$

or in the simply connected case

$$\{\text{parallel spinors}\} \simeq V_{\tilde{\mathfrak{h}}} := \left\{ v \in \Delta_{r,s} \middle| (\Phi_{(r,s)})_*(\tilde{\mathfrak{h}})(v) = 0 \right\}.$$

Now one wants to find properties of manifolds admitting parallel spinors. On has the following

1.16 Proposition. *([Hit74] for the Riemannian case, [Bau81] for the general) Let (M,h) be a connected semi-Riemannian spin manifold with a parallel spinor field. Then the Ricci-endomorphism is totally isotropic, i.e. $h(Ric(X), Ric(Y)) = 0$ for all $X, Y \in TM$.*

Now one assigns to every spinor field a vector field via a inner product on S.

1.17 Definition. Let (M, h) be a semi-Riemannian manifold of index (r, s) and $\varphi \in \Gamma(S)$. The vector field $V_\varphi \in \Gamma(TM)$ defined by $h(V_\varphi, X) = i^{r+1}\langle X\cdot\varphi, \varphi\rangle$ for all $X \in TM$ is called **φ-associated vector field.**
If (M, h) is a Lorentzian manifold then V_φ is called **Dirac-current** and satisfies $h(V_\varphi, X) = -\langle X \cdot \varphi, \varphi\rangle$.

Now a parallel spinor entails a parallel associated vector field.

1.18 Lemma. *Let (M, h) be a semi-Riemannian manifold and $\varphi \in \Gamma(S)$ a parallel spinor. Then V_φ is parallel or zero.*

Proof. From the relations between the covariant derivatives, the Clifford multiplication and the scalar product it is for all $X, Y \in TM$

$$\begin{aligned} h\left(\nabla_Y V_\varphi, X\right) &= Y\left(h\left(V_\varphi, X\right)\right) - h\left(V_\varphi, \nabla_Y X\right) \\ &= i^{r+1}\left(Y\left(\langle X\cdot\varphi,\varphi\rangle\right) - \langle\nabla_Y X\cdot\varphi,\varphi\rangle\right) \\ &= i^{r+1}\left(\langle\nabla_Y X\cdot\varphi,\varphi\rangle + \langle X\cdot\nabla^S_Y\varphi,\varphi\rangle + \langle X\cdot\varphi,\nabla^S_Y\varphi\rangle - \langle\nabla_Y X\cdot\varphi,\varphi\rangle\right) \\ &= 0 \ \text{ if } \varphi \text{ parallel.} \end{aligned}$$

This gives the conclusion. □

Clearly V_φ can be zero, as it is the case for Riemannian manifolds. Here the existence of a parallel spinor does not imply the existence of a non-trivial parallel vector field. But for Lorentzian manifolds we will see that the zeros of the Dirac current are equal to the zeros of the spinor field.

1.2.3 Parallel spinors on product manifolds

Here we will prove that a product of two semi-Riemannian manifolds carries parallel spinors if and only if each of the factors does. For this we need a lemma.

1.19 Lemma. *Let ρ be a representation of a Lie group $G := G_1 \times G_2$ and ρ_i representations of G_i such that $\rho \sim \rho_1 \otimes \rho_2$. Then ρ has a trivial subrepresentation if and only if both ρ_i have a trivial sub-representation.*

Proof. One direction is obvious: If $\Psi : V \to V_1 \otimes V_2$ is the isomorphism between the representation spaces and v_i fixed vectors under ρ_i $(i = 1, 2)$ then $v := \Psi^{-1}(v_1 \otimes v_2)$ is fixed under ρ.
Otherwise let v be a fixed vector under ρ. Then $\Psi(v)$ is a fixed vector under $\rho_1 \otimes \rho_2$, i. e. it defines a trivial representation of $G_1 \times G_2$ which is one-dimensional and thus irreducible. A theorem in representation theory (see for example [Tit67]) says that every

irreducible representation of a group product $G_1 \times G_2$ splits into a tensor product of two irreducible representations of G_1 and G_2. Hence $\mathbb{K}\ \Psi(v) = U_1 \otimes U_2$ where U_1 and U_2 are one-dimensional, of course. Thus $\Psi(v) = \sum_{j=1}^k u_1^j \otimes u_2^j = \sum_{j=1}^k a^j b^j\ u_1 \otimes u_2$ with scalars a^j, b^j and $u_1 \in U_1$, $u_2 \in U_2$ since U_1 and U_2 are one dimensional. But this implies

$$u_1 \otimes u_2 = \rho_1(g_1) \otimes \rho_2(g_2)\,(u_1 \otimes u_2) = \rho_1(g_1)u_1 \otimes \rho_2(g_2)u_2.$$

for any pair $(g_1, g_2) \in G_1 \times G_2$. Setting $g_1 = 1$ or $g_2 = 1$ implies $\rho_i(g_i)u_i = u_i$ for all $g_j \in G_i$, $i = 1, 2$. □

With the help of this lemma we get:

1.20 Proposition. *Let be $H_1 \subset SO(r_1, s_1) \subset Gl(m_1, \mathbb{R})$ and $H_2 \subset SO(r_2, s_2) \subset Gl(m_2, \mathbb{R})$ with $r_1 + s_1 = m_1$ and $r_2 + s_2 = m_2$. Let*

$$H := H_1 \times H_2 \ \subset\ SO(r_1, s_1) \times SO(r_2, s_2) \ \subset\ SO(r_1 + r_2, s_1 + s_2) \ \subset\ Gl(m, \mathbb{R})$$

be the usual inclusion with $m = m_1 + m_2$. Then it holds: The spin representation of $H \subset SO(r_1 + r_2, s_1 + s_2)$ has a trivial subrepresentation if and only if the spin representations of $H_1 \subset SO(r_1, s_1)$ and $H_2 \subset SO(r_2, s_2)$ have trivial subrepresentations.

Proof. Let H_1, H_2 and H be as in the assumption of the theorem. We denote the corresponding lifts in the spin group via the twofold covering λ by $\tilde{H}, \tilde{H}_1$ and $\tilde{H}_2$. Then we have the following commuting diagram

$$\begin{array}{ccccc}
\tilde{H} = \left(\tilde{H}_1 \times \tilde{H}_2\right)_{/\mathbb{Z}_2} & \subset & (Spin(r_1, s_1) \times Spin(r_2, s_2))_{/\mathbb{Z}_2} & \hookrightarrow & Spin(r, s) \\
 & & \downarrow \lambda \times \lambda & & \downarrow \lambda \\
H = H_1 \times H_2 & \subset & O(r_1, s_1) \times O(r_2, s_2) & \hookrightarrow & O(r, s),
\end{array}$$

where the imbeddings of the product groups into the $O(r, s)$ resp. $Spin(r, s)$ come from the canonical identification of $\mathbb{R}^{m_1} \times \mathbb{R}^{m_2}$ and $\mathbb{R}^m$ sending $e_i \in \mathbb{R}^{m_1}$ onto $e_i \in \mathbb{R}^m$ and $e_i \in \mathbb{R}^{m_2}$ onto $e_{m_1+i} \in \mathbb{R}^m$.
Hence we have to show that the spin representation $\Phi_{(r,s)}\big|_{\tilde{H}}$ of $\tilde{H} = (\tilde{H}_1 \times \tilde{H}_2)_{/\mathbb{Z}_2}$ has a trivial subrepresentation if and only if the spin representation $\Phi_{(r_1,s_1)}\big|_{\tilde{H}_1}$ and $\Phi_{(r_2,s_2)}\big|_{\tilde{H}_2}$ of $\tilde{H}_1$ and $\tilde{H}_1$ have trivial subrepresentations. The latter is the case — by the lemma — if and only if the representation $\left(\Phi_{(r_1,s_1)} \otimes \Phi_{(r_2,s_2)}\right)\big|_{(\tilde{H}_1 \times \tilde{H}_2)_{/\mathbb{Z}_2}}$ has a trivial subrepresentation.
To get this, we consider two representations of $(Spin(r_1, s_1) \times Spin(r_2, s_2))_{/\mathbb{Z}_2}$: The first is given as the representation tensor product:

$$\begin{array}{rcccl}
\pi & : & (Spin(r_1, s_1) \times Spin(r_2, s_2))_{/\mathbb{Z}_2} & \to & End(\Delta_{r_1,s_1} \otimes \Delta_{r_2,s_2}) \\
 & & [(A, B)] & \mapsto & \Phi_{(r_1,s_1)}(A) \otimes \Phi_{(r_2,s_2)}(B).
\end{array}$$

This definition is correct since $\pi([(A,B)]) = \pi([(-A,-B)])$.
The second results from the restriction of $\Phi_{(r,s)}$ on $Spin(r_1,s_1) \times (Spin(r_2,s_2) \hookrightarrow Spin(r,s))_{/\mathbb{Z}_2}$:

$$\Phi := \Phi_{(r,s)}\Big|_{(Spin(r_1,s_1)\times Spin(r_2,s_2))/\mathbb{Z}_2}$$

If m is even the latter has the two irreducible subrepresentations $\Phi^{\pm}_{(r,s)} : Spin(r,s) \to End(\Delta^{\pm}_{r,s})$ which can be restricted again. The restricted representation we denote by $\Phi^{\pm}$. With this notations we get:

1.21 Lemma. *For the defined representations of $Spin(r_1,s_1) \times Spin(r_2,s_2)$ holds the following*

1. *If m_1 or m_2 is even then we have $\Phi \sim \pi$.*

2. *For both m_1 and m_2 odd we have $\Phi^{\pm} \sim \pi$.*

Proof. 1.) Let $m_1 = 2\,k_1$ be even and $m = 2k$ in case m_2 being even too and $m = 2k+1$ in case m_2 being odd. The isomorphism between the representation spaces $\Delta_{r,s}$ and $\Delta_{r_2,s_2} \otimes \Delta_{r_1,s_1}$ is given by the identity in both cases, denoted by:

$$\begin{array}{rccl} \Psi : & \Delta_{r,s} & \to & \Delta_{r_2,s_2} \otimes \Delta_{r_1,s_1} \\ & u(\varepsilon_k,\ldots,\varepsilon_1) & \mapsto & u(\varepsilon_k,\ldots,\varepsilon_{k_1+1}) \otimes u(\varepsilon_{k_1},\ldots,\varepsilon_1) \end{array}$$

For the Clifford-multiplication, that is for $\Phi_{(r,s)}$ as the representation of the Clifford algebra we have for $X \in \mathbb{R}^{m_1}$ and the corresponding $\hat{X} \in \mathbb{R}^m$:

$$\begin{aligned} \Psi\left(\hat{X}\cdot u(\varepsilon_k,\ldots,\varepsilon_1)\right) &= \Psi\left(\Phi_{(r,s)}\left(\hat{X}\right)(u(\varepsilon_k,\ldots,\varepsilon_1))\right) \\ &= u(\varepsilon_k,\ldots,\varepsilon_{k_1+1}) \otimes \Phi_{(r_1,s_1)}(X)\,(u(\varepsilon_{k_1},\ldots,\varepsilon_1)) \\ &= u(\varepsilon_k,\ldots,\varepsilon_{k_1+1}) \otimes X\cdot u(\varepsilon_{k_1},\ldots,\varepsilon_1) \end{aligned}$$

and for $Y \in \mathbb{R}^{m_2}$ and the corresponding $\hat{Y} \in \mathbb{R}^m$:

$$\begin{aligned} &\Psi\left(\hat{Y}\cdot u(\varepsilon_k,\ldots,\varepsilon_1)\right) \\ &= \Psi\left(\Phi_{(r,s)}\left(\hat{Y}\right)(u(\varepsilon_k,\ldots,\varepsilon_1))\right) \\ &= (-1)^{k_1}\varepsilon_1\cdot\ldots\cdot\varepsilon_{k_1}\ \Phi_{(r_2,s_2)}(Y)\,(u(\varepsilon_k,\ldots,\varepsilon_{k_1+1}))\otimes u(\varepsilon_{k_1},\ldots,\varepsilon_1) \\ &= (-1)^{k_1}\varepsilon_1\cdot\ldots\cdot\varepsilon_{k_1}\ Y\cdot u(\varepsilon_k,\ldots,\varepsilon_{k_1+1})\otimes u(\varepsilon_{k_1},\ldots,\varepsilon_1). \end{aligned}$$

Therefore we have for $g = Y_1\cdot\ldots\cdot Y_{2i} \in Spin(r_2,s_2)$ with $Y_l \in \mathbb{R}^{m_2}$ and $h = X_1\cdot\ldots\cdot X_{2j} \in Spin(r_1,s_1)$ with $X_l \in \mathbb{R}^{m_1}$

$$\begin{aligned} (\pi(g,h)\circ\Psi)(u(\varepsilon_k,\ldots,\varepsilon_1)) &= \Phi_{(r_2,s_2)}(g)u(\varepsilon_k,\ldots,\varepsilon_{k_1+1})\otimes\Phi_{(r_1,s_1)}(h)u(\varepsilon_{k_1},\ldots,\varepsilon_1) \\ &= \Psi\left(\Phi_{(r,s)}(g\cdot h)\,(u(\varepsilon_k,\ldots,\varepsilon_1))\right) \\ &= \Psi\circ\Phi(g,h)\,(u(\varepsilon_k,\ldots,\varepsilon_1))\,, \end{aligned}$$

which shows that both representations are equivalent.
2.) Let now be $m_i = 2\,k_i + 1$ such that $m = m_1 + m_2 = 2\,(k+1)$ with $k = k_1 + k_2$. For technical reasons we consider a $\iota \in O(r,s)$ which is the identity on $span(e_1, \dots, e_{m_1})$ and further defined by

$$\iota(e_{m_1+i}) = e_{m_1+1+i} \text{ for } 1 \leq i < m_2 \quad \text{and} \quad \iota(e_m) = e_{m_1+1}.$$

The resulting representation of the Clifford algebra $\Phi_\iota := \Phi_{(r,s)} \circ \iota$ is equivalent to $\Phi_{(r,s)}$. This is true, because the definition of the twofold covering λ allows us to write $\iota = \lambda(g)$ for $g \in Spin(r,s)$ and therefore

$$\Phi_\iota(X) = \Phi_{(r,s)}(\lambda(g)(X)) = \Phi_{(r,s)}(g \cdot X \cdot g^{-1}) = g \cdot \Phi_{(r,s)}(X) \cdot g^{-1}.$$

This verifies the equivalence.
Now we will show that the corresponding irreducible representation $\Phi_\iota^\pm$ are equivalent to π.
The representation space of $\Phi_\iota^\pm$ is $\mathbb{C}^{2^k} \simeq \Delta_{r,s}^\pm \subset \Delta_{r,s} \simeq \mathbb{C}^{2^{k+1}}$ and of π is $\Delta_{r_2,s_2} \otimes \Delta_{r_1,s_1} \simeq \mathbb{C}^{2^k}$. The isomorphism is the following

$$\begin{aligned} \Psi^\pm : \Delta_{r,s}^\pm &\to \Delta_{r_1,s_1} \otimes \Delta_{r_2,s_2} \\ u(\varepsilon_{k+1}, \dots, \varepsilon_1) &\mapsto a_{\varepsilon_{k+1},\dots,\varepsilon_{k_1+2}} u(\varepsilon_{k+1}, \dots, \varepsilon_{k_1+2}) \otimes u(\varepsilon_{k_1}, \dots, \varepsilon_1) \\ &\quad \text{with } a_{\varepsilon_{k+1},\dots,\varepsilon_{k_1+2}} = \begin{cases} 1 & : \prod_{l=1}^{k_2} \varepsilon_{k_1+1+l} = 1 \\ i\,(-1)^{k_2+1} & : \prod_{l=1}^{k_2} \varepsilon_{k_1+1+l} = -1 \end{cases} \end{aligned}$$

One has to verify that $\Psi^\pm \circ \Phi_\iota^\pm(g,h) = \pi(g,h) \circ \Psi^\pm$. First we show this relation for g and h of the form $e_i \cdot e_j$ with $e_i, e_j \in \mathbb{R}^{m_2}$ resp. $\mathbb{R}^{m_1}$.
For $i = j$ this relation holds clearly since e_i^2 is a multiplication with a scalar.
We distinguish the following cases:

1. $e_i, e_j \in \mathbb{R}^{m_1}$ and

   (a) both different from e_{m_1}. Then it is clear that

   $$\begin{aligned} &\Psi^\pm(\iota(e_i) \cdot \iota(e_j) \cdot u(\varepsilon_{k+1}, \dots, \varepsilon_1)) \\ =\ &\Psi^\pm(u(\varepsilon_{k+1}, \dots, \varepsilon_{k_1+1}) \otimes e_i \cdot e_j \cdot u(\varepsilon_{k_1}, \dots, \varepsilon_1)) \\ =\ &\pi(id, e_i \cdot e_j) \circ \Psi^\pm(u(\varepsilon_k, \dots, \varepsilon_1)). \end{aligned}$$

   (b) one of them equals to e_{m_1}. W.l.o.g. should be $j = m_1$. Then it is

$$\begin{aligned} &\Psi^\pm(\iota(e_i) \cdot e_{m_1} \cdot u(\varepsilon_{k+1}, \dots, \varepsilon_1)) \\ =\ &\Psi^\pm\Big(\iota(e_i) \cdot (-1)^{k_1} \varepsilon_1 \cdot \ldots \cdot \varepsilon_{k_1}\ i\ u(\varepsilon_{k+1}, \dots, \varepsilon_{k_1+2}, -\varepsilon_{k_1+1}, \varepsilon_{k_1}, \dots, \varepsilon_1)\Big) \\ =\ &\Psi^\pm\Big((-1)^{k_1} \varepsilon_1 \cdot \ldots \cdot \varepsilon_{k_1}\ i\ u(\varepsilon_{k+1}, \dots, \varepsilon_{k_1+2}, -\varepsilon_{k_1+1}) \otimes e_i \cdot u(\varepsilon_{k_1}, \dots, \varepsilon_1)\Big) \\ =\ &\Psi^\pm(u(\varepsilon_{k+1}, \dots, -\varepsilon_{k_1+1}) \otimes e_i \cdot e_{m_1} \cdot u(\varepsilon_{k_1}, \dots, \varepsilon_1)) \\ =\ &\pi(id, e_i \cdot e_{m_1}) \circ \Psi^\pm(u(\varepsilon_k, \dots, \varepsilon_1)). \end{aligned}$$

So one has for $e_i, e_j \in \mathbb{R}^{m_1}$

$$\Psi^{\pm} \circ \Phi^{\pm}_{\iota}(id, e_i \cdot e_j) = \pi(id, e_i \cdot e_j) \circ \Psi^{\pm}$$

and therefore for all $h \in Spin(r_1, s_1)$

$$\Psi^{\pm} \circ \Phi^{\pm}_{\iota}(id, h) = \pi(id, h) \circ \Psi^{\pm}.$$

2. $e_i, e_j \in \mathbb{R}^{m_2}$ and

   (a) both not equal to e_{m_2}. Here we have again

$$\begin{aligned}
&\Psi^{\pm}\left(\iota(e_i) \cdot \iota(e_j) \cdot u(\varepsilon_{k+1}, \ldots, \varepsilon_1)\right) \\
&= \Psi^{\pm}\left(e_{m_1+1+i} \cdot e_{m_1+1+j} \cdot u(\varepsilon_{k+1}, \ldots, \varepsilon_1)\right) \\
&= (-1)^{2(k_1+1)} \Psi^{\pm}\left((e_i \cdot e_j \cdot u(\varepsilon_{k+1}, \ldots, \varepsilon_{k_1+2})) \otimes u(\varepsilon_{k_1+1}, \ldots, \varepsilon_1)\right) \\
&= \pi(e_i \cdot e_j, id) \circ \Psi^{\pm} u(\varepsilon_k, \ldots, \varepsilon_1).
\end{aligned}$$

   (b) $e_j = e_{m_2}$. Then

$$\begin{aligned}
&\Psi^{\pm}\left(\iota(e_i) \cdot \iota(e_{m_2}) \cdot u(\varepsilon_{k+1}, \ldots, \varepsilon_1)\right) \\
&= \Psi^{\pm}\left(e_{m_1+1+i} \cdot e_{m_1+1} \cdot u(\varepsilon_{k+1}, \ldots, \varepsilon_1)\right) \\
&= \Psi^{\pm}\left((e_i \cdot u(\varepsilon_{k+1}, \ldots, \varepsilon_{k_1+2})) \otimes u(-\varepsilon_{k_1+1}, \varepsilon_{k_1}, \ldots, \varepsilon_1)\right) \\
&= a_{\varepsilon_{k+1}, \ldots, \varepsilon_{k_1+2}} \left(e_{m_2} \cdot e_i \cdot u(\varepsilon_{k+1}, \ldots, \varepsilon_{k_1+2})\right) \otimes u(\varepsilon_{k_1}, \ldots, \varepsilon_1) \\
&= \pi(e_{m_2} \cdot e_i, id) \circ \Psi^{\pm}\left(u(\varepsilon_{k+1}, \ldots, \varepsilon_1)\right).
\end{aligned}$$

So one has for all $g \in Spin(r_2, s_2)$ that

$$\Psi^{\pm} \circ \Phi^{\pm}_{\iota}(g, id) = \pi(g, id) \circ \Psi^{\pm}$$

and therefore the conclusion of the lemma. □

By this equivalence of representations one concludes: $\Phi_{(r_1,s_1)}\big|_{\tilde{H}_1}$ and $\Phi_{(r_2,s_2)}\big|_{\tilde{H}_2}$ of $\tilde{H}_1$ and $\tilde{H}_1$ have trivial subrepresentations if and only if $\pi\big|_{(\tilde{H}_1 \times \tilde{H}_2)/\mathbb{Z}_2}$ has a trivial subrepresentation if and only if $\phi\big|_{(\tilde{H}_1 \times \tilde{H}_2)/\mathbb{Z}_2}$ resp. $\phi^{\pm}\big|_{(\tilde{H}_1 \times \tilde{H}_2)/\mathbb{Z}_2}$ has a trivial subrepresentation. But this gives the result. □

Now let (M_i, h_i) be two semi-Riemannian manifolds of dimension m_i and signature (r_i, s_i) and (M, h) the product of both of signature $(r = r_1 + r_2, s = s_1 + s_2)$. Its holonomy is given by the product

$$H := Hol_{(x_1,x_2)}(M, h) = Hol_{x_1}(M_1, h_1) \times Hol_{x_2}(M_2, h_2) =: H_1 \times H_2$$

For these we get as a consequence:

1.22 Corollary. *On a semi-Riemannian manifold product manifold*

$$(M,h) \simeq (M_1,h_1) \times (M_2,h_2)$$

there exist parallel spinors if and only if on both semi-Riemannian manifolds (M_i,h_i) *exist parallel spinors.*

1.3 Decomposition of a Lorentzian manifold with parallel spinors

From now on let (M,h) be a Lorentzian manifold. For Lorentzian manifolds, the zeros of the Dirac current correspond to the zeros of the spinor field. Hence we can use the Dirac current of a parallel spinor field to decompose the manifold.
For the Dirac-current the following properties are valid.

1.23 Lemma. *Let* (M,h) *a Lorentzian spin manifold and* $\varphi \in \Gamma(S)$. *Then* V_φ *is causal, i.e.* $h(V_\varphi, V_\varphi) \leq 0$. *Furthermore:* $V_\varphi(x) = 0$ *if and only if* $\varphi(x) = 0$.

Proof. First of all the definition entails that the zeros of the spinor field are zeros of the vector field.
Since (M,h) is spin we have ξ the timelike global defined unit vector field. Then we have

$$h(V_\varphi,\xi) = -\langle \xi \cdot \varphi, \varphi\rangle := -(\xi\cdot\xi\cdot\varphi,\varphi)_\xi = -(\varphi,\varphi)_\xi \neq 0 \text{ if } \varphi \neq 0.$$

This implies that the zeros of the vector field are zeros of the spinor field. Furthermore one can write $V_\varphi = (\varphi,\varphi)_\xi\, \xi + \eta$ with a space like vector field η.
Now we consider an arbitrary $x \in M$. If $\eta(x) = 0$ then $V_\varphi(x)$ is timelike and we are done.
Let therefore $\eta(x) \neq 0$. Then on sets

$$s_1 := \xi \text{ and } s_2 := \frac{1}{\sqrt{h(\eta,\eta)}}\eta$$

and completes this locally with vector fields $s_3, \ldots, s_m$ of length 1 to a local orthonormal basis. Since $h(V_\varphi, X) = -\langle X \cdot \varphi, \varphi\rangle$ we have in this basis

$$V_\varphi = -\sum_{i=1}^{n} \kappa_i \langle s_i \cdot \varphi, \varphi\rangle s_i = (\varphi,\varphi)_\xi\; s_1 - (s_1 \cdot s_2 \cdot \varphi, \varphi)_\xi\, s_2$$

and for the length $h(V_\varphi, V_\varphi) = -(\varphi,\varphi)_\xi^2 + (s_1 \cdot s_2 \cdot \varphi, \varphi)_\xi^2$.
The action of $e_1 \cdot e_2$ on the basis of the spinor module ($k := [\frac{m}{2}]$) is

$$e_1 \cdot e_2 \cdot u(\varepsilon_k, \ldots, \varepsilon_1) = -\varepsilon_1 u(\varepsilon_k, \ldots, \varepsilon_1).$$

Thus for $\varphi \in S$ with $\varphi(x) = [\hat{s}(x), v(x)] \in \widetilde{\mathcal{O}}_\xi \times_K \Delta_{1,m-1}$ where $v(x) \in \mathbb{C}^k$ we have $(\varphi(x), \varphi(x))_\xi = (v(x), v(x))$ and $(s_1 \cdot s_2 \cdot \varphi(x), \varphi(x))_\xi = (e_1 \cdot e_2 \cdot v(x), v(x))$ and we decompose $v(x)$ as follows

$$
\begin{aligned}
v(x) &= \sum_{\varepsilon_1,\dots,\varepsilon_k=\pm 1}^{k} v_{(\varepsilon_k,\dots,\varepsilon_1)} u\left(\varepsilon_k,\dots,\varepsilon_1\right) \\
&= \sum_{\varepsilon_2,\dots,\varepsilon_k=\pm 1}^{k} v_{(\varepsilon_k,\dots,\varepsilon_2,1)} u\left(\varepsilon_k,\dots,\varepsilon_2,1\right) + \sum_{\varepsilon_2,\dots,\varepsilon_k=\pm 1}^{k} v_{(\varepsilon_k,\dots,\varepsilon_2,-1)} u\left(\varepsilon_k,\dots,\varepsilon_2,-1\right) \\
&=: v_+ + v_- .
\end{aligned}
$$

So we have $e_1 \cdot e_2 \cdot v(x) = -v_+ + v_-$ and for the length

$$
\begin{aligned}
(e_1 \cdot e_2 \cdot v(x), v(x)) &= (-v_+ + v_-, v_+ + v_-) \\
&= -\left(v_+, v_+\right) + \left(v_-, v_-\right) - \underbrace{\left(v_+, v_-\right)}_{=0} + \underbrace{\left(v_-, v_+\right)}_{=0} \\
&= -\left(v_+, v_+\right) + \left(v_-, v_-\right).
\end{aligned}
$$

$$
\begin{aligned}
\text{Finally:}\quad h\left(V_\varphi, V_\varphi\right) &= -\left(\varphi,\varphi\right)^2 + \left(s_1 \cdot s_2 \cdot \varphi, \varphi\right)^2 \\
&\overset{x}{=} -\Big(\left(v_+, v_+\right)^2 + 2\ \left(v_+, v_+\right)\left(v_-, v_-\right) + \left(v_-, v_-\right)^2\Big) \\
&\quad + \Big(\left(v_+, v_+\right)^2 - 2\ \left(v_+, v_+\right)\left(v_-, v_-\right) + \left(v_-, v_-\right)^2\Big) \\
&= -4\ \left(v_+, v_+\right)\left(v_-, v_-\right) \leq 0.
\end{aligned}
$$

Hence V_φ is causal. □

From the second assertion of the lemma follows that a non-flat Lorentzian manifold with parallel spinors cannot be irreducible as we have seen in corollary 1.14.
The second point suggests to distinguish two cases for a **simply connected, complete Lorentzian manifold with parallel spinor** φ and associated parallel Dirac current V_φ:
1.) $h\left(V_\varphi, V_\varphi\right) < 0$, i.e. V_φ timelike. Since V_φ is parallel the holonomy group acts trivial on $\mathbb{R}\ V_\varphi(x) \subset T_xM$, and so the manifold decomposes due to theorem 1.12 as follows

$$(M,h) \simeq (\mathbb{R}, -dt) \times (N,g),$$

with (N,g) a Riemannian manifold of dimension $m-1$ with parallel spinor field (due to corollary 1.22.) The latter can be decomposed in flat or irreducible Riemannian manifolds, again with parallel spinors.
Since irreducible symmetric Riemannian manifolds with parallel spinors has to be flat (as a conclusion of the Ricci-flatness, see [Bes87]) non of these irreducible factors is symmetric. Since all of the factors are complete, non of them can be locally symmetric. Hence all the factors has to be entries of the list of Wang.

1.24 Theorem. *(M, h) is a simply connected, complete Lorentzian manifold with parallel spinor whose associated vector field not lightlike if and only if (M, h) is isometric to a product of $(\mathbb{R}, -dt^2)$ and simply connected, complete irreducible Riemannian manifolds with one of the following holonomy groups $SU(k)$, $Sp(k)$, G_2, $Spin(7)$ and eventually a flat factor.*

2.) Let V_φ lightlike. I.e. the holonomy representation has an one-dimensional, degenerate, trivial sub-representation. The de-Rham decomposition yields a proper indecomposable, i.e. non-irreducible factor, which has to be Lorentzian of course. Hence (M, h) decomposes in a product of simply connected, complete, irreducible Riemannian manifolds with parallel spinors and a simply connected, complete, indecomposable Lorentzian manifold which is non-irreducible and which admits in addition a parallel spinor field with parallel lightlike Dirac current.
In the following chapter we will collect some properties of the holonomy of simply connected, compete Lorentzian manifold which are indecomposable, but non-irreducible.

Chapter 2

Indecomposable Lorentzian manifolds

This chapter deals with indecomposable, non-irreducible Lorentzian manifolds of dimension $n+2$. First of all we describe the basic algebraic properties of such a holonomy group. Its algebra is contained in $(\mathbb{R} \oplus \mathfrak{so}(n)) \ltimes \mathbb{R}^n$ We introduce the notion of the *screen bundle* and show that the $\mathfrak{so}(n)$-projection can be understood as holonomy algebra of the connection on the screen bundle induced by the Levi-Civita connection. Furthermore we describe the results of L. Berard-Bergery and A. Ikemakhen. The second section deals with the local description of an indecomposable, non-irreducible Lorentzian manifold, its distributions, foliations and local coordinates. We describe the results of C. Boubel. In the last section sufficient conditions for the existence of manifolds with special kind of holonomy are given.

Throughout the whole chapter we suppose that (M, h) is a simply connected $n+2$–dimensional Lorentzian manifold, such that the full holonomy equals to the reduced one, which is connected. The following facts are true also for the reduced holonomy of a Lorentzian manifold which is not simply connected.

2.1 The holonomy of an indecomposable Lorentzian manifold

2.1.1 Basic properties

Let (M, h) be a simply connected Lorentzian manifold of dimension $m = n + 2 \geq 3$. We consider the tangent bundle as associated to the fibre bundle $\mathcal{L}(M, h)$ with fibres over $p \in M$

$$\mathcal{L}_p(M,h) := \left\{ (X, s_1, \dots s_n, Z) \,\middle|\, \begin{array}{l} \text{a basis in } T_pM \text{ with } h(X,Z) = 1, \\ h(X,X) = h(Z,Z) = h(X,s_i) = h(Z,s_i) = 0, \\ h(s_i,s_j) = \delta_{ij} \end{array} \right\}$$

and structure group $O(\mathbb{R}^{n+2},\eta) := \{A \in GL(n+2,\mathbb{R}) | A^t\eta A = \eta\}$ where η is defined as follows: $\eta := \begin{pmatrix} 0 & 0^t & 1 \\ 0 & E_n & 0 \\ 1 & 0^t & 0 \end{pmatrix}$.

Now we suppose that (M,h) is indecomposable but non-irreducible, not necessarily with a fixed lightlike vector under holonomy representation but only with an invariant degenerate subspace $\mathbb{E} \subset T_pM$. Then $\mathbb{E}^\perp$ is invariant and defines an invariant, isotropic, one-dimensional subspace $\mathbb{X} := \mathbb{E} \cap \mathbb{E}^\perp \neq \{0\}$. This subspace defines via theorem 1.10 an one-dimensional isotropic parallel distribution Ξ. Hence we have a reduction of $\mathcal{L}(M,h)$ and of the Levi-Civita connection to the bundle

$$\mathcal{P}(M,h) := \{(X, s_1, \dots s_n, Z) \in \mathcal{L}(M,h) | X \in \Xi\} \tag{2.1}$$

and structure group $P = \mathrm{Iso}(\mathbb{R}e_0) \subset O(\mathbb{R}^{n+2},\eta)$ acting on $\mathbb{R}^{n+2} = span(e_0,\dots,e_{n+1})$. This groups is often called parabolic group.

Algebraic remarks P can be written as follows:

$$P = \left\{ \begin{pmatrix} a & -av^tA & -\frac{a}{2}v^tv \\ 0 & A & v \\ 0 & 0^t & a^{-1} \end{pmatrix} \,\middle|\, a \in \mathbb{R}^*, v \in \mathbb{R}^n, A \in O(n) \right\}. \tag{2.2}$$

It has the following closed subgroups

$$\mathbb{R}^* \simeq \left\{ \begin{pmatrix} a & 0^t & 0 \\ 0 & E_n & 0 \\ 0 & 0^t & a^{-1} \end{pmatrix} \,\middle|\, a \in \mathbb{R}^* \right\} \text{ and } O(n) \simeq \left\{ \begin{pmatrix} 1 & 0^t & 0 \\ 0 & A & 0 \\ 0 & 0^t & 1 \end{pmatrix} \,\middle|\, A \in O(n) \right\}, \tag{2.3}$$

which commute with each other, and the closed vector group

$$\mathbb{R}^n \simeq \left\{ \begin{pmatrix} 1 & v^t & -\frac{1}{2}v^tv \\ 0 & E_n & -v \\ 0 & 0^t & 1 \end{pmatrix} \,\middle|\, v \in \mathbb{R}^n \right\}, \tag{2.4}$$

which is normal in P. Furthermore it is $(\mathbb{R}^* \times O(n)) \cap \mathbb{R}^n = \{1\}$, and every element in P can be written as product of an element in $\mathbb{R}^* \times O(n)$ and an element in $\mathbb{R}^n$. This is equivalent to the fact that P is a semi-direct product

$$P = (\mathbb{R}^* \times O(n)) \ltimes \mathbb{R}^n \qquad \text{with respect to the action of } \mathbb{R}^* \times O(n) \text{ on } \mathbb{R}^n$$
$$(a, A) : \mathbb{R}^n \ni v \mapsto (a, A) \cdot v \cdot (a^{-1}, A^{-1}) = aAv \in \mathbb{R}^n.$$

The Lie algebra of P — written in matrices — is

$$\mathfrak{p} = \left\{ \left(\begin{array}{ccc} a & v^t & 0 \\ 0 & A & -v \\ 0 & 0^t & -a \end{array} \right) \middle| \, a \in \mathbb{R}, v \in \mathbb{R}^n, A \in \mathfrak{so}(n) \right\} \subset \mathfrak{so}(\mathbb{R}^{n+2}, \eta). \tag{2.5}$$

It is isomorphic to $\mathbb{R} \oplus \mathfrak{so}(n) \oplus \mathbb{R}^n$ as vector space with the commutator:

$$[(a, A, v), (b, B, w)] = \left(0, [A, B]_{\mathfrak{so}(n)}, (A + a\, Id)\, w - (B + b\, Id)\, v\right). \tag{2.6}$$

In this sense we will refer to $\mathbb{R}$, $\mathbb{R}^n$ and $\mathfrak{so}(n)$ as subalgebras of $\mathfrak{p} \subset \mathfrak{so}(\mathbb{R}^{n+2}, \eta)$. $\mathbb{R}$ is an abelian subalgebra of $\mathfrak{p}$, $\mathbb{R}^n$ an abelian ideal in $\mathfrak{p}$, and it holds

$$[\mathbb{R}, \mathfrak{so}(n)] = 0 \quad [\mathbb{R}, \mathbb{R}^n] \subset \mathbb{R}^n \quad [\mathfrak{so}(n), \mathbb{R}^n] \subset \mathbb{R}^n. \tag{2.7}$$

We have $\mathfrak{p} = (\mathbb{R} \oplus \mathfrak{so}(n)) \ltimes \mathbb{R}^n$.
Now one can assign to a subalgebra $\mathfrak{h} \subset \mathfrak{p}$ the projections $pr_{\mathbb{R}}(\mathfrak{h})$, $pr_{\mathbb{R}^n}(\mathfrak{h})$ and $pr_{\mathfrak{so}(n)}(\mathfrak{h})$. Then the following properties are obvious.

2.1 Proposition. *Let $\mathfrak{h}$ be an indecomposably acting subalgebra of $\mathfrak{p}$ and set $\mathfrak{g} := pr_{\mathfrak{so}(n)}(\mathfrak{h})$. Then holds*

1. *$\mathfrak{h}$ has no other invariant subspace then the defining one, which is fixed by $\mathfrak{p}$.*
2. *$pr_{\mathbb{R}^n}(\mathfrak{h}) = \mathbb{R}^n$.*
3. *$\mathfrak{h}$ has a trivial subrepresentation if and only if $pr_{\mathbb{R}}\mathfrak{h} = 0$.*
4. *$\mathfrak{h}$ is abelian if and only if $\mathfrak{h} = \mathbb{R}^n$.*
5. *$\mathfrak{g}$ is compact and therefore reductive.*

A Lie algebra **compact** if there exists a positive definite invariant symmetric bilinear form on it. Subalgebras of compact algebras are compact. A Lie algebra $\mathfrak{g}$ is called **reductive** if its Levi decomposition is $\mathfrak{g} = \mathfrak{z} \oplus \mathfrak{d}$ where $\mathfrak{z}$ is the center of $\mathfrak{g}$ and $\mathfrak{d} := [\mathfrak{g}, \mathfrak{g}]$ the derived Lie algebra, which is is semisimple. The invariant, positive definite symmetric bilinear form of a compact Lie algebra gives a Levi decomposition into the center and the derived Lie algebra, since $[\mathfrak{g}, \mathfrak{g}]^{\perp} = \mathfrak{z}$. Hence compact Lie algebras are reductive. All together implies the last point.

We now return to the group P as structure group of an indecomposable, non-irreducible Lorentzian manifold. Since the manifold is orientable the holonomy group of the latter is contained in the connection component of the unit $P_0 = (\mathbb{R}^+ \times SO(n)) \ltimes \mathbb{R}^n$ in P, and defined up to conjugation in P. Choosing a different basis of T_pM in $\mathcal{P}_p(M,h)$ and changing the point p corresponds to conjugation in P.
Thus the projections of $\mathfrak{h} := \mathfrak{hol}_p(M,h)$ on $\mathbb{R}$, $\mathbb{R}^n$ and $\mathfrak{so}(n)$ are defined up to conjugation in P. But the projection on $\mathbb{R}$ is independent of conjugation, since it is zero or one-dimensional. Also the projection on $\mathbb{R}^n$ is independent because it is the whole $\mathbb{R}^n$. From the commutator relation (2.6) follows, that $pr_{\mathfrak{so}(n)}$ is a Lie algebra homomorphism from $\mathfrak{p}$ to $\mathfrak{so}(n)$. Hence the $\mathfrak{so}(n)$–projection of $\mathfrak{hol}_p(M,h)$ is defined up to conjugation in $O(n)$ and represented on $\mathbb{R}^n$ in the standard way.

To characterize the geometric properties of an indecomposable Lorentzian manifold we recall the following definition.

2.2 Definition. A vector field X is called **recurrent** if $\nabla X = \theta \otimes X$ with a 1–form θ. A Lorentzian manifold with a lightlike parallel vector field is called **Brinkmann-wave**. (See [Bri25], also [Eis38].)

Then theorem 1.10 implies the following:

2.3 Proposition. *Let $\mathfrak{h} := \mathfrak{hol}_p(M,h)$ be the indecomposable, but non-irreducible holonomy algebra of a Lorentzian manifold (M,h). Then there exists a recurrent, lightlike vector field X on M. It is parallel (i.e. (M,h) is a Brinkmann-wave) if and only if* $pr_{\mathbb{R}}\mathfrak{h} = 0$.

2.1.2 The $\mathfrak{so}(n)$-projection as holonomy group

In this section we will describe the $SO(n)$–projection of an indecomposable, non-irreducible holonomy group of a $n+2$–dimensional, simply connected Lorentzian manifold as a holonomy group of a metrical connection in a vector bundle, the so called "screen bundle".
Again we consider distributions Ξ and $\Xi^\perp$ on M, which are parallel, i. e. $\nabla_U : \Gamma(\Xi^{(\perp)}) \to \Gamma(\Xi^{(\perp)})$ for every $U \in TM$. In every point $p \in M$ one considers the factor space $\Xi_p^\perp/\Xi_p$. This defines a vector bundle

$$E := \dot{\bigcup_{p\in M}} \Xi_p^\perp/\Xi_p \longrightarrow M$$
$$[U_p] \longmapsto p.$$

This vector bundle is called **screen bundle**. Now the metric h on M defines a scalar product on E, which we denote by $\hat{h}$, via

$$\hat{h}\left([X],[Y]\right) := h(X,Y).$$

With respect to this scalar product the bundle $\mathcal{O}(E)$ is defined as the set of orthonormal frames of E over M. This is a $O(n)$–principal fibre bundle. $\mathcal{O}(E)$ has fibres

$$\mathcal{O}_p(E) = \left\{ ([s_1], \dots, [s_n]) \;\middle|\; \begin{array}{l} (X, s_1, \dots, s_n) \text{ a basis of } \Xi_p^\perp \text{ for a } 0 \neq X \in \Xi_p \\ \text{with } h(s_i, s_j) = \delta_{ij} \end{array} \right\}.$$

Then we can describe E as vector bundle associated to the bundle $\mathcal{O}(E)$:

$$\begin{array}{rcl} \mathcal{O}(E) \times_{O(n)} \mathbb{R}^n & \simeq & E \\ \left[([s_1], \dots, [s_n]), (x_1, \dots x_n)\right] & \mapsto & \left[\sum_{i=1}^n x_i s_i\right] \end{array}$$

We now consider again the principal fibre bundle $\mathcal{P}(M, h)$ with structure group P from the beginning of this chapter in formula (2.1). We define a surjective bundle homomorphism

$$\begin{array}{rccc} f : & \mathcal{P}(M, h) & \to & \mathcal{O}(E) \\ & (X, s_1, \dots, s_n, Z) & \mapsto & ([s_1], \dots, [s_n]). \end{array}$$

Then holds the following:

2.4 Lemma. *$f : \mathcal{P}(M, h) \to \mathcal{O}(E)$ is a $pr_{O(n)} : P \to O(n)$–reduction*

Proof. We have to verify that the following diagram commutes

$$\begin{array}{ccccc} P \times \mathcal{P}(M, h) & \longrightarrow & \mathcal{P}(M, h) & & \\ & & & \searrow & \\ pr_{SO(n)} \times f \downarrow & \circlearrowleft & f \downarrow & \circlearrowleft & M. \\ & & & \nearrow & \\ O(n) \times \mathcal{O}(E) & \longrightarrow & \mathcal{O}(E) & & \end{array}$$

It is $P = (\mathbb{R}^* \times O(n)) \ltimes \mathbb{R}^n$ with $\mathbb{R}^*$ as in (2.3), $\mathbb{R}^n$ as in (2.4) and $O(n) \subset O(\eta)$. Then we have

$$(X, s_1, \dots, s_n, Z) \cdot (a, Id, 0) = (aX, s_1, \dots, s_n, a^{-1} Z) \tag{2.8}$$

and

$$\begin{array}{l} (X, s_1, \dots, s_n, Z) \cdot (1, Id, v) = \\ (X, v_1 X + s_1, \dots, v_n X + s_n, -\frac{1}{2} v^t v \, X - \sum_{k=1}^n v_k s_k + Z). \end{array} \tag{2.9}$$

Since P is a semi-direct product this implies that

$$f\left((X, s_1, \dots, s_n, Z) \cdot (a, A, v)\right) = ([s_1], \dots, [s_n]) \cdot A.$$

But this makes the left part of the above diagram commutative. For the right part this is obvious. □

Since Ξ is parallel the Levi-Civita connection defines also a covariant derivative ∇^E on E by

$$\nabla^E_X[Y] := [\nabla_X Y].$$

This covariant derivative is metrical with respect to $\hat{h}$ since the Levi-Civita connection is metrical. It defines a connection form θ on $\mathcal{O}(E)$ which is given for a local section $\hat{\sigma} = ([\sigma_1], \ldots, [\sigma_n]) \in \Gamma(\mathcal{O}(E))$ by the formula

$$\nabla^E_U[V] = \nabla^E_U[(\hat{\sigma}, \nu)] = \left[\hat{\sigma}, d\nu(V) + \theta^{\hat{\sigma}}(U) \cdot \nu\right]$$

for $\nu = (\nu_1, \ldots, \nu_n)$ and $[V] = \sum_{i=1}^n \nu_i[\sigma_i]$ locally, where $\theta^{\hat{\sigma}}$ is the local connection form of θ. We get the following

2.5 Proposition. *Let (M, h) be an indecomposable Lorentzian manifold of dimension $n+2$ and with parallel distribution Ξ. Let ω denote the connection form of the Levi-Civita connection ∇. Then ω is a $pr_{O(n)}$-reduction of the connection θ of $\mathcal{O}(E)$.*

Proof. We consider the diagram

$$\begin{array}{ccc} T\mathcal{P}(M,h) & \xrightarrow{df} & T\mathcal{O}(E) \\ \omega\downarrow & & \downarrow\theta \\ \mathfrak{p} & \xrightarrow[dpr_{O(n)}=pr_{\mathfrak{so}(n)}]{} & \mathfrak{so}(n) \end{array} \qquad (2.10)$$

and have to show that $(df)_s$ sends the kernel of ω_s to the kernel of $\theta_{f(s)}$ for $s \in \mathcal{P}(M, h)$. Now every element in the kernel of ω_s is equal to $(d\sigma)_p(U)$ for $p \in M$, $U \in T_pM$ and a certain local section $\sigma \in \Gamma(\mathcal{P}(M, h))$ with $\sigma(p) = s$. Now it is

$$0 = \omega_{\sigma(p)}((d\sigma)_p(U)) = (\sigma^*\omega)_p(U) = \omega^\sigma_p(U).$$

For the local connection form ω^σ of the Levi-Civita connection one calculates as follows: For $\sigma = (\xi, \sigma_1, \ldots, \sigma_n, \zeta) \in \Gamma(\mathcal{P}(M, h))$ and E_{rt} the standard basis of $\mathfrak{gl}(n, \mathbb{R})$ it is

$$\begin{array}{rcll} 0 = \omega^\sigma(U) & = & h(\nabla_U \xi, \zeta)\,(E_{00} - E_{n+1n+1}) & \text{(the } \mathbb{R}\text{–part)} \\ & & + \sum_{k=1}^{n} h(\nabla_U \sigma_k, \zeta)\,(E_{0k} - E_{kn+1}) & \text{(the } \mathbb{R}^n\text{–part)} \\ & & + \sum_{1\le k<l\le n} h(\nabla_U \sigma_k, \sigma_l)\,(E_{kl} - E_{lk}) & \text{(the } \mathfrak{so}(n)\text{–part).} \end{array}$$

We have to consider $(df)_{\sigma(p)}(d\sigma)_p(U) = d(f \circ \sigma)_p(U)$. If now $\sigma \in \Gamma(\mathcal{P}(M, h))$ as above, then is $f \circ \sigma = ([\sigma_1], \ldots, [\sigma_n]) \in \Gamma(\mathcal{O}(E))$. Finally it is

$$\begin{aligned} \theta_{f\circ\sigma(p)}(d(f \circ \sigma)_p(U)) &= \theta^{f\circ\sigma}(U) \\ &= \sum_{1\le k<l\le n} \hat{h}(\nabla^E_U[\sigma_k], [\sigma_l])\,(E_{kl} - E_{lk}) \\ &= \sum_{1\le k<l\le n} h(\nabla^E_U \sigma_k, \sigma_l)\,(E_{kl} - E_{lk}) \\ &= 0 \end{aligned}$$

because of the above equation. I.e. $d(f \circ \sigma)_p(U)$ is in the kernel of the local connection $\theta^{f \circ \sigma}$. So it is in the kernel of θ. □

2.6 Corollary. *The diagram (2.10) commutes and for the curvatures Θ of θ and Ω of ω holds*

$$f^*\Theta = pr_{\mathfrak{so}(n)} \circ \Omega.$$

This implies the following for the holonomy algebras:

$$\mathfrak{hol}_p(E, \nabla^E) = pr_{\mathfrak{so}(n)}(\mathfrak{hol}_p(M, h)).$$

Proof. By the fact that f and df are surjective and the Ambrose-Singer holonomy theorem one gets $\mathfrak{hol}_{f(s)}(\theta) = pr_{\mathfrak{so}(n)}(\mathfrak{hol}_s(\omega))$, and this gives the last assertion. □

Unfortunately there is no holonomy classification of metrical covariant derivatives in vector bundles which are not tangent bundles. We cannot use this result to a classification of possible $SO(n)$–projections of Lorentzian holonomy groups.
But this description can be used to interprete the geometric information encoded in $\mathfrak{g} := pr_{\mathfrak{so}(n)}(\mathfrak{hol}(M, h))$ as geometric structure on the screen bundle E. For example if there is a complex structure on E which is compatible with the metric $\hat{h}$ and parallel to the covariant derivative ∇^E, which was induced by the Levi-Civita connection of (M, h), then the flag $\Xi \subset \Xi^\perp \subset TM$ is called **Kähler flag**. The existence of such a Kähler flag is equivalent to $\mathfrak{g} \subset \mathfrak{u}(n)$. For $\mathfrak{g} \subset \mathfrak{su}(n)$ one calls such a flag **special Kähler flag**. For details see [Bau03] and [Kat99].This can be done analogously for any other geometric structure on E, resp. $\mathfrak{g}$.

2.1.3 Results of A. Ikemakhen and L. Berard-Bergery

The most important results about indecomposable Lorentzian holonomy are due to A. Ikemakhen and L. Berard-Bergery [BI93]. They studied all three projections of an indecomposable Lorentzian holonomy algebra. For $\mathfrak{g} := pr_{\mathfrak{so}(n)}(\mathfrak{hol}_p(M, h))$ they proved a so-called Borel-Lichnerowicz property. Furthermore they described the relations between the three projections of an indecomposable subalgebra of $\mathfrak{p}$.

Borel-Lichnerowicz property

2.7 Theorem. *[BI93] Let $\mathfrak{g} := pr_{\mathfrak{so}(n)}\left(\mathfrak{hol}_p(M, h)\right)$ the projection of the holonomy algebra of an indecomposable, non-irreducible Lorentzian manifold onto the $\mathfrak{so}(n)$-component. Then $\mathfrak{g}$ is completely reducible and there exists decompositions of $\mathbb{R}^n$ in orthogonal subspaces and of $\mathfrak{g} \subset \mathfrak{so}(n)$ in ideals*

$$\mathbb{R}^n = E_0 \oplus E_1 \oplus \ldots \oplus E_r \quad \textit{and} \quad \mathfrak{g} = \mathfrak{g}_1 \oplus \ldots \oplus \mathfrak{g}_r$$

where $\mathfrak{g}$ acts trivial on E_0, $\mathfrak{g}_i$ acts irreducible on E_i and $\mathfrak{g}_i(E_j) = \{0\}$ for $i \neq j$.

To draw a first conclusion we have to cite another general fact about subgroups of $SO(n)$.

2.8 Proposition. *Let G be connected.*

1. *If G is semisimple and $\mathfrak{g}$ compact, then G is compact (H. Weyl, see for example [Bou82]).*
2. *If $G \subset SO(n)$ acts irreducible on $\mathbb{R}^n$, then the center is discrete or isomorphic to S^1 (see for example [KN63, Appendix 5]).*

In particular holds that irreducibly acting, connected subgroups of $SO(n)$ are compact.

2.9 Corollary. *Let (M,h) be an indecomposable, non-irreducible, simply connected Lorentzian manifold. Then $G = pr_{SO(n)}(Hol_p(M,h))$ is compact.*

But we will see that the whole holonomy group is not closed in general.

Four algebraic types The second result of [BI93] is the distinction of indecomposable subalgebras of $\mathfrak{p}$ into four types due to the relation between their projections. For this result — in contrary to theorem 2.7 — it is not supposed that the subalgebra is a Lorentzian holonomy algebra.

2.10 Theorem. *[BI93] Let $\mathfrak{h}$ be a subalgebra of $\mathfrak{p} \subset \mathfrak{so}(\mathbb{R}^{n+2}, \eta)$ which acts indecomposable on $\mathbb{R}^{n+2}$, $\mathfrak{g} := pr_{\mathfrak{so}(n)}(\mathfrak{h}) = \mathfrak{z} \oplus \mathfrak{d}$ as above. Then $\mathfrak{h}$ belongs to one of the following types.*

1. *If $\mathfrak{h}$ contains $\mathbb{R}^n$, then we have the types*

   Type 1: *$\mathfrak{h}$ contains $\mathbb{R}$. Then $\mathfrak{h} = (\mathbb{R} \oplus \mathfrak{g}) \ltimes \mathbb{R}^n$.*

   Type 2: *$pr_{\mathbb{R}}(\mathfrak{h}) = 0$ i.e. $\mathfrak{h} = \mathfrak{g} \ltimes \mathbb{R}^n$.*

   Type 3: *Neither Type 1 nor Type 2.*

   In that case there exists a surjective homomorphism $\varphi : \mathfrak{z} \to \mathbb{R}$, such that

   $$\mathfrak{h} = (\mathfrak{l} \oplus \mathfrak{d}) \ltimes \mathbb{R}^n$$

   where $\mathfrak{l} := graph\ \varphi = \{(\varphi(T), T) | T \in \mathfrak{z}\} \subset \mathbb{R} \oplus \mathfrak{z}$. Or written as matrices:

   $$\mathfrak{h} = \left\{ \left. \begin{pmatrix} \varphi(A) & v^t & 0 \\ 0 & A+B & -v \\ 0 & 0 & -\varphi(A) \end{pmatrix} \right| A \in \mathfrak{z}, B \in \mathfrak{d}, v \in \mathbb{R}^n \right\}.$$

2. *In case $\mathfrak{h}$ does not contain $\mathbb{R}^n$ we have* **Type 4:**

   There exists

(a) *a non-trivial decomposition* $\mathbb{R}^n = \mathbb{R}^k \oplus \mathbb{R}^l$, $0 < k, l < n$,

(b) *a surjective homomorphism* $\varphi : \mathfrak{z} \to \mathbb{R}^l$

such that $\mathfrak{g} \subset \mathfrak{so}(k)$ *and* $\mathfrak{h} = (\mathfrak{d} \oplus \mathfrak{l}) \ltimes \mathbb{R}^k \subset \mathfrak{p}$ *where* $\mathfrak{l} := \{(\varphi(T), T) \, | T \in \mathfrak{z}\} =$ *graph* $\varphi \subset \mathbb{R}^l \oplus \mathfrak{z}$. *Or written as matrices:*

$$\mathfrak{h} = \left\{ \left. \begin{pmatrix} 0 & \varphi(A)^t & v^t & 0 \\ 0 & 0 & A+B & -v \\ 0 & 0 & 0 & -\varphi(A) \\ 0 & 0 & 0 & 0 \end{pmatrix} \right| A \in \mathfrak{z}, B \in \mathfrak{d}, v \in \mathbb{R}^k \right\}.$$

These four algebraic types give four types of indecomposable Lorentzian holonomy with the same properties. Of course these types are independent of conjugation in P.
We denote by Latin capitals the corresponding connected Lie groups to the Lie algebras of the theorem. Then D is compact, because it is semisimple and $\mathfrak{d}$ is compact (Theorem of H. Weyl). G is compact because of corollary 2.9.
The group L is closed if and only if the kernel of φ generates a compact subgroup of Z.

2.11 Corollary. *[BI93] Let H be the holonomy group of a simply connected, indecomposable, non-irreducible Lorentzian manifold. Then H is of one of the four types of indecomposable subalgebras of the parabolic algebra $\mathfrak{p}$ and it holds for*

Type 1: $H = (\mathbb{R}^* \times G) \ltimes \mathbb{R}^n$ *is closed.*

Type 2: $H = G \ltimes \mathbb{R}^n$ *closed.*

Type 3: $H = (L \times D) \ltimes \mathbb{R}^n$ *is closed if and only if $Ker\ \varphi$ generates a compact subgroup of the center of G. The center of G is non-trivial.*

Type 4: $H = (L \times D) \ltimes \mathbb{R}^k$ *is closed if and only if $Ker\ \varphi$ generates a compact subgroup of the center of G. Furthermore the non-triviality of the decomposition of $\mathbb{R}^n$ forces the center of G to be non-trivial.*

2.12 Corollary. *An indecomposable, reducible Lorentzian manifold is of type 2 or 4 if and only if it is a Brinkmann-wave.*

2.13 Remark. 1. In [BI93] for every of these types examples of metrics are constructed. In particular metrics whose holonomy is of type 3 and 4 and non-closed. The latter are constructed with the help of the dense immersion of the real line in the torus T^2 which plays the role of the center of G.

2. In the following section we will how the $\mathfrak{so}(n)$–projection is related to a Riemannian holonomy algebra, and why it is sensible to ask, whether it is equal to a

Riemannian holonomy algebra. The question, if there is an indecomposable, non-irreducible Lorentzian holonomy, whose $SO(n)$-projection is not a Riemannian holonomy will occupy further sections of the present thesis.

2.1.4 14 holonomy types of space-time

It is possible to classify all possible holonomy groups of 4-dimensional Lorentzian manifolds, i.e. of space times in general relativity. For sake of completeness we will present the result here. It is due to [Sch60] and [Sha70] and rediscovered by [Hal93] and [HL00]. [Sch60] and [Sha70] classified all possible subgroups of $SO(1,3)$. If they are indecomposable we will them corresponding to the four types of indecomposable subalgebras of the parabolic algebras due to Berard-Bergery and Ikemakhen:

1. irreducible: $SO(1,3)$
2. indecomposable of type 1: $(\mathbb{R} \times SO(2)) \ltimes \mathbb{R}^2$
3. indecomposable of type 2: $SO(2) \ltimes \mathbb{R}^2$.
4. indecomposable of type 3: $graph\ \varphi \ltimes \mathbb{R}^2$ with $\varphi : SO(2) \to \mathbb{R}$,
5. decomposable, but not completely reducible: $A_1 \times 1$ with A_1 subgroup of type 1 of $SO(1,2)$, isomorphic to $\mathbb{R}^2$
6. decomposable, but not completely reducible: $A_2 \times 1$ with A_2 subgroup of type 2 of $SO(1,2)$, isomorphic to $\mathbb{R}$
7. completely reducible: $1 \times SO(3)$
8. completely reducible: $1 \times 1 \times SO(2)$
9. completely reducible: $SO(1,2) \times 1$
10. completely reducible: $SO(1,1) \times SO(2)$
11. completely reducible: an one-dimensional subgroup of $SO(1,1) \times SO(2)$ with non-trivial parts in both summands,
12. $\{1\}$.

The case 11 cannot occur because of the decomposition theorem of Wu. Beside this it is possible to embed $SO(2)$ into $SO(3)$ in three possible ways. Hence we get 14 possible holonomy groups of 4-dimensional Lorentz manifolds. The case of an indecomposable type 4 algebra cannot occur for dimensional reasons.

2.2 Local description of an indecomposable Lorentzian manifold

In this section we will recall some results about the local shape of indecomposable Lorentzian manifolds. Based on foliations of the manifold which are defined by the holonomy invariant subspaces, we will give different coordinates and the local form of the metric. These coordinates are due to [Bri25], [Wal49] and [Bou00].

2.2.1 Parallel distributions and foliations

Parallel distributions The holonomy invariant, lightlike subspace $\mathbb{X}$ and its orthogonal complement $\mathbb{X}^\perp$ in a tangent space are fibres of distributions

$$\Xi \subset \Xi^\perp \ , \ \Xi \text{ lightlike.} \tag{2.11}$$

These are parallel, so they are involutive and therefore integrable. For every point $p \in M$, there are integral manifolds $\mathcal{X}_p$ and $\mathcal{X}_p^\perp$ of Ξ and $\Xi^\perp$ passing through it.
For a $p \in M$ we have $\Xi_p \subset \Xi_p^\perp \subset T_pM$ and we can consider the space

$$\hat{\Xi}_p^\perp := \Xi_p^\perp / \Xi_p \qquad \text{with the canonical projection } \Xi_p^\perp \to \hat{\Xi}_p^\perp,$$
$$\text{and positive definite scalar product } \hat{h}_p([X],[Y]) = h_p(X,Y).$$

Let $\mathfrak{h} := \mathfrak{hol}_p(M,h)$ be the holonomy algebra acting on T_pM. Since Ξ_p and $\Xi_p^\perp$ are invariant, this action defines a $\hat{h}_p$-orthogonal action of $\mathfrak{h}$ on $\hat{\Xi}_p^\perp$ via $A[X] = [AX]$. If we denote this Lie algebra by $\hat{\mathfrak{h}}$, then we have

$$\hat{\mathfrak{h}} \subset \mathfrak{so}(\hat{\Xi}_p^\perp, \hat{h}_p).$$

But this implies, that $\hat{\mathfrak{h}}$ acts completely reducibly on $\hat{\Xi}_p^\perp$, i.e. there is an $\hat{h}$-orthogonal and $\hat{\mathfrak{h}}$-invariant decomposition

$$\hat{\Xi}_p^\perp = \hat{\Upsilon}_p^0 \oplus \ldots \oplus \hat{\Upsilon}_p^s$$

in subspaces on which $\hat{\mathfrak{h}}$ acts trivial — $\hat{\Upsilon}_p^0$ — or irreducible — $\hat{\Upsilon}_p^1, \ldots, \hat{\Upsilon}_p^s$.
Let the spaces Υ_p^i be the pre-image of $\hat{\Upsilon}_p^i$ with intersection Ξ_p. These are holonomy invariant. Therefore they are the fibres of parallel distributions $\Upsilon^0, \ldots, \Upsilon^s$ on M with

$$\Xi = \Upsilon^0 \cap \ldots \cap \Upsilon^s. \tag{2.12}$$

In this context hold the following

2.14 Lemma. *[Bou00] Let $X \in \Xi, Y^i \in \Upsilon^i$ and $Y \in \Xi^\perp$. Then holds for the curvature $\mathcal{R}$ of (M,h)*

1. $\mathcal{R}(Y^i, Y^j) = \mathcal{R}(Y^0, Y) = \mathcal{R}(X, Y) = 0$ *for* $i \neq j$.

2. *For* $U, V \in TM$ *and* $i \neq j$ *holds* $\mathcal{R}(Y^i, U, Y^j, V) = \mathcal{R}(Y^i, V, Y^j, U)$.

Foliations and coordinates on the leaves The Frobenius theorem furthermore implies that all these distributions define foliations on M, denoted by $\mathcal{X}$, $\mathcal{Y}^i$ and $\mathcal{X}^\perp$, — a fact which is used in the next section — but also on the leaves $\mathcal{X}_p^\perp$. Each leave of $\mathcal{Y}^i$ and $\mathcal{X}^\perp$ again is foliated in leaves of $\mathcal{X}$.

The leaves of $\mathcal{X}$ are a lightlike geodesic lines, the leaves of $\mathcal{Y}^i$ are lightlike totally geodesic submanifolds, and the ones of $\mathcal{X}^\perp$ are lightlike totally geodesic hypersurfaces. Thus for every leaf $\mathcal{X}_p^\perp$ exists different types of coordinate systems (U,φ) around $p \in M$, all with $\varphi(p) = 0$.

First we have of course coordinates of $\mathcal{X}_p^\perp$ adapted to the foliation $\mathcal{X} \subset \mathcal{X}^\perp$, i.e. coordinates of the form

$$(U, \varphi = (x, y_1, \ldots, y_n)) \tag{2.13}$$

such that the leaves of $\mathcal{X}$ are parameterized by x and with the property

$$\nabla_{\frac{\partial}{\partial y_k}} \frac{\partial}{\partial x} = 0.$$

In $\mathcal{X}_p^\perp$ these coordinates define n–dimensional submanifolds

$$W_x := \{\varphi^{-1}(x, y_1, \ldots y_n) | (y_1, \ldots y_n) \in \mathbb{R}^n \cap \varphi(U)\}$$

The restriction of the metric h to its tangent spaces is positive definite, so these can be considered as Riemannian manifolds. The Riemannian metric will be denoted by $\hat{h}^x$ and corresponds to the above $\hat{h}$.

With the help of lemma 2.14 one can prove the existence of special coordinates on the leaves.

2.15 Lemma. *[Bou00] There are coordinates adapted to the foliation $\mathcal{X} \subset \mathcal{Y}^0, \ldots, \mathcal{Y}^s \subset \mathcal{X}^\perp$, i.e. of the form*

$$\left(U, \left(x, y_1^0, \ldots, y_{d_0}^0, \ldots, y_1^s, \ldots, y_{d_s}^s\right)\right) \tag{2.14}$$

with $d_i = dim \hat{\Upsilon}^i$ and the following properties:

- *x parameterizes the leaves of $\mathcal{X}$ and $(x, y_1^i, \ldots, y_{d_i}^i)$ the leaves of $\mathcal{Y}^i$ for $i = 0, \ldots, s$,*
- *$\nabla_{\frac{\partial}{\partial y_k^i}} \frac{\partial}{\partial x} = \nabla_{\frac{\partial}{\partial y_k^i}} \frac{\partial}{\partial y_l^j} = 0$ for $i \neq j$ and $(\nabla_{\frac{\partial}{\partial y_k^0}})|_{T\mathcal{X}_p^\perp} = 0$,*
- *$h_p\left(\frac{\partial}{\partial y_k^i}(p), \frac{\partial}{\partial y_l^j}(p)\right) = \delta_{ij}\delta_{kl}$.*

Again these coordinates define Riemannian submanifolds W^i with metrics $\hat{h}^i$ depending on the parameters x and y_l^j for all $j \neq i$. With (W_0^i, h_0^i) we denote the Riemannian manifold with all parameters zero, i.e. containing p.

Local sections Here we will illustrate how the choice of these coordinates corresponds to a section of the fibration $\pi : \mathcal{X}_p^\perp \to (\mathcal{X}_p^\perp)/\mathcal{X} := \hat{\mathcal{X}}_p^\perp$ for a fixed $p \in M$. This fibration gives a commuting diagram

$$\begin{array}{ccc} T\mathcal{X}_p^\perp = \Xi^\perp_{|\mathcal{X}_p^\perp} & \stackrel{d\pi}{\longrightarrow} & T\hat{\mathcal{X}}_p^\perp. \\ \downarrow & \circlearrowleft & \downarrow \\ \mathcal{X}_p^\perp & \stackrel{\pi}{\longrightarrow} & \hat{\mathcal{X}}_p^\perp. \end{array}$$

Then for every $q \in \mathcal{X}_p^\perp$ holds that

$$T_{\pi(q)}\hat{\mathcal{X}}_p^\perp \simeq \Xi_q^\perp / Ker(d\pi)_q = \Xi_q^\perp/\Xi_q = \hat{\Xi}_q^\perp. \tag{2.15}$$

Now we consider a section $\sigma \in \Gamma(\mathcal{X}_p^\perp \to \hat{\mathcal{X}}_p^\perp)$ of π. It defines a Riemannian metric $g^\sigma := \sigma^* h$ on $\hat{\mathcal{X}}_p^\perp$. For the Levi-Civita connection ∇^σ to this metric one verifies with the help of the Koszul formula that

$$g^\sigma\left(\nabla^\sigma_U V, W\right) = h\left(\nabla_{\sigma_*(U)}\sigma_* V, \sigma_* W\right) \quad \text{i.e.} \quad \sigma_*(\nabla^\sigma_U V) = \nabla_{\sigma_* U}\sigma_* V.$$

where ∇ is the Levi-Civita connection of h. But this implies the following for the parallel displacements $\mathcal{P}$ of ∇ and $\mathcal{P}^\sigma$ of ∇^σ:

$$(d\pi)_{\sigma\circ\gamma(1)} \circ \mathcal{P}_{\sigma\circ\gamma} = \mathcal{P}^\sigma_\gamma \circ (d\pi)_{\sigma(\gamma(0))}.$$

With the help of the above identification (2.15) one gets now

$$\mathfrak{hol}_{\pi(p)}(\hat{\mathcal{X}}_p, g^\sigma) \subset \hat{\mathfrak{h}}.$$

This result was first proved by A. Ikemakhen [Ike96] in terms of local coordinates.
The above decomposition of the tangent space

$$T_{\hat{q}}\hat{\mathcal{X}} \simeq \hat{\Xi}^\perp_{\sigma(\hat{q})} = \hat{\Upsilon}^0_{\sigma(\hat{q})} \oplus \ldots \oplus \hat{\Upsilon}^s_{\sigma(\hat{q})}$$

is $\hat{\mathfrak{h}}$– and therefore $\mathfrak{hol}_{\hat{q}}(g^\sigma)$)–invariant. If now the section σ satisfies the property that its differential restricted to $\hat{\Upsilon}^i_q$ is a section of $\Upsilon^i_q \to \hat{\Upsilon}^i_q$, i.e.

$$d\sigma_{|\hat{\Upsilon}^i_q} \in \Gamma(\Upsilon^i_q \to \hat{\Upsilon}^i_q)$$

for all $q \in \mathcal{X}_p^\perp$, $i = 0, \ldots, s$, then this decomposition is also g^σ orthogonal, i.e. the distributions Υ^i are ∇^σ–parallel and g^σ–orthogonal on $\hat{\mathcal{X}}^\perp$, and define a decomposition

$$(\hat{\mathcal{X}}_p^\perp, g^\sigma) \stackrel{\text{isometric}}{\simeq} (\hat{\mathcal{Y}}^0_p, g^0) \times \ldots \times \hat{\mathcal{Y}}^s_p, g^s)$$

where the metrics g^i are the restrictions of the metric g^σ on $T\hat{\mathcal{Y}}^i \simeq \hat{\Upsilon}^i_{\sigma(.)}$. These Riemannian manifolds are not necessarily irreducible, since the inclusion $\mathfrak{hol}_{\pi(p)}(\hat{\mathcal{X}}_p, g^\sigma) \subset \hat{\mathfrak{h}}$ is not necessary irreducible.
Coordinates on the Riemannian manifolds $\mathcal{Y}^i_p$ now can be lifted via σ to coordinates of $\mathcal{X}_p^\perp$. These define again Riemannian manifolds $(W^i, \hat{h}^i)$ which are isometric to $(\hat{\mathcal{Y}}^i, g^i)$.

2.2.2 Local coordinates

These foliations and coordinates yield coordinates of the manifold (M, h) with certain conditions. This can be seen by considering the foliation $\mathcal{X}^\perp \subset M$. Then a section of the fibration $M \to M/\mathcal{X}^\perp$ defines a coordinate z by lifting the coordinate of the 1-dimensional manifold $M/\mathcal{X}^\perp$. Now one has to prolongate the above coordinates on one leaf $\mathcal{X}_p^\perp$ along the coordinate lines of z, maintaining the "good" properties on the leaf. C. Boubel showed the conditions for doing this uniquely [Bou00].

Walker coordinates These coordinates are adapted to the foliation $\mathcal{X} \subset \mathcal{X}^\perp \subset M$.

2.16 Proposition. *1. [Wal49], [Ike96] (M, h) is a Lorentzian manifold of dimension $n + 2 > 2$ with recurrent, lightlike vector field if and only if there exists coordinates $(U, \varphi = (x, (y_i)_{i=1}^n, z))$ in which the metric h has the following local shape*

$$h = 2\, dxdz + \sum_{i=1}^{n} u_i dy_i dz + f dz^2 + \sum_{i,j=1}^{n} g_{ij} dy_i\, dy_j \tag{2.16}$$

with $\frac{\partial g_{ij}}{\partial x} = \frac{\partial u_i}{\partial x} = 0$, $f \in C^\infty(M)$. If f is sufficient general ($\frac{\partial f}{\partial y_i} \neq 0$ for all $i = 1, \dots, n$), then (M, h) is indecomposable.

2. *[Bri25][Eis38] The existence of a parallel lightlike vector field is equivalent to the additional condition $\frac{\partial f}{\partial x} = 0$.*

3. *[Sch74] (M, h) is a Lorentzian manifold with parallel lightlike vector field if and only if there exists coordinates $(U, \varphi = (x, (y_i)_{i=1}^n, z))$ in which h has the shape*

$$h = 2\, dxdz + \sum_{i,j=1}^{n} g_{ij} dy_i\, dy_j \ , \text{ with } \frac{\partial g_{ij}}{\partial x} = 0. \tag{2.17}$$

It is clear that the vector field $\frac{\partial}{\partial x}$ corresponds to the recurrent/parallel lightlike vector field.

If one considers small n-dimensional submanifolds in U through $p = \varphi^{-1}(x, y_1, \dots, y_n, z)$ defined by

$$W_{(x,z)} := \{\varphi^{-1}(x, y_1, \dots y_n, z) | (y_1, \dots y_n) \in \mathbb{R}^n \cap \varphi(U)\}$$

then one can understand the g_{ij} as coefficients of a family of Riemannian metrics g_z and the u_i as coefficients of a family of 1-forms ϕ_z on $W_{(x,z)}$ depending on a parameter z.

Adapted coordinates with uniqueness condition Here we start with coordinates on the leaves of $\mathcal{X}^{\perp}$ adapted to the foliations $\mathcal{Y}^i$. With the help of lemma 2.15 one proves the following theorem, in which γ plays the role of ϕ, not considered as a family of 1–forms, but as 1–form on the whole coordinate neighborhood.

2.17 Theorem. *[Bou00] Let (M,h) be an indecomposable Lorentzian manifold of dimension $n+2>2$ with a recurrent lightlike vector field. Then there exists coordinates*

$$(U,\varphi=(x,y_1^0,\dots,y_{d_0}^0,\dots,y_1^s,\dots,y_{d_s}^s,z))$$

around the point $p\in M$, such that

$$h=2\;dxdz+\gamma\;dz+\sum_{i=0}^{s}\sum_{k,l=1}^{d_i}g_{kl}^i dy_k^i dy_l^i, \tag{2.18}$$

which are adapted to the foliations $(\mathcal{X},\mathcal{Y}^0,\dots,\mathcal{Y}^s,\mathcal{X}^{\perp})$, i.e.

- *$g_{kl}^i\in C^{\infty}(M)$ with $\frac{\partial}{\partial x}(g_{kl}^i)=\frac{\partial}{\partial y_m^j}(g_{kl}^i)=0$ for $i\neq j$ and*
- *γ a 1-form with $d\gamma(\frac{\partial}{\partial x},\frac{\partial}{\partial y_l^i})=d\gamma(\frac{\partial}{\partial y_k^i},\frac{\partial}{\partial y_l^j})=0$ for $i\neq j$.*

Furthermore these coordinates can be chosen in a way such that

1. *$g_{kl}^0=\delta_{kl}$ and $\gamma(\frac{\partial}{\partial z})=0$,*
2. *the initial condition $g_{kl}^i(p)=\delta_{kl}$ holds,*
3. *γ satisfies:*
   (a) *On $\mathcal{X}_p^{\perp}$ holds $\gamma=dx$.*
   (b) *On the curve μ_z with the coordinates $(0,\dots,0,z)$ holds $\gamma=dx$, and the 1-form $\gamma_{|\Xi_{\mu_z}^{\perp}}$ is closed.*
   (c) *Let $\mathcal{S}_z^i:=\{q\in\mathcal{Y}_{\mu_z}^i|x(q)=0\}$. Then holds for all $q\in\mathcal{S}_z^i$, that $\gamma_{|T_q\mathcal{S}_z^i}=0$.*

Adapted coordinates with 1, 2 and 3 are uniquely determined by its values on the initial manifold $\mathcal{X}_p^{\perp}$.
The 1-form γ is uniquely determined by the three conditions 3a, 3b and 3c and the relation

$$\gamma\left(\frac{\partial}{\partial x}\right)\frac{\partial}{\partial z}\left(\frac{\partial}{\partial y_k^i}(\gamma(Y))\,\gamma(\frac{\partial}{\partial x})^{-1}\right)=\mathcal{R}\left(\frac{\partial}{\partial z},\frac{\partial}{\partial y_k^i},\frac{\partial}{\partial z},Y\right) \tag{2.19}$$

for $Y=\frac{\partial}{\partial y_l^j}$ with $i\neq j$ or $i=0$ or $Y=\frac{\partial}{\partial x}$.

2.18 Remark. In case that $\hat{\mathfrak{h}}$ acts irreducible, the unique form of (2.18) corresponds to the form(2.17).

In [Bou00] theorem 2.17 is used to find equivalent conditions for an indecomposable, non-irreducible Lorentzian manifold to have holonomy of type 1,2,3 or 4. We will cite it here only for the coupled type 3. For type 4 holds an analogous result.

2.19 Theorem. *[Bou00] Let (M,h) be an indecomposable Lorentzian manifold of dimension $n+2>2$ with a recurrent lightlike vector field. Let $p\in M$ and*

$$(U,\varphi=(x,y_1^0,\dots,y_{d_0}^0,\dots,y_1^s,\dots,y_{d_s}^s,z))$$

coordinates around p due to theorem 2.17. Then the holonomy of (M,h) is of type 3 if and only if it is of type 1 or 3 there is the following decomposition

$$\hat{\Xi}_p^{\perp}=\hat{\Upsilon}_p^0\oplus\hat{\Upsilon}_p^1\oplus\dots\oplus\hat{\Upsilon}_p^r\oplus\hat{\Upsilon}_p^{r+1}\oplus\dots\oplus\hat{\Upsilon}_p^s$$

with $r\geq 1$ and

1. *$\hat{\mathfrak{h}}$ acts trivial on $\hat{\Upsilon}_p^0$.*

2. *For $i=1,\dots,r$ and for every $q\in U$ the leaves $\hat{\mathcal{Y}}_q^i$ are Ricci flat Kähler manifolds of dimension d^i with complex structure J^i and $\hat{\mathfrak{h}}_{|\hat{\Upsilon}_q^i}$ equals to $\mathbb{R}J^i\oplus\mathfrak{s}^i$ where $\mathfrak{s}^i$ is an irreducibly acting subalgebra of $\mathfrak{su}\left(\hat{\Upsilon}_q^i,\hat{h}_q^i,J^i\right)$.*

3. *There is a $\lambda\in\mathbb{R}^r$ such that in every $q\in U$ holds*

$$\gamma(\frac{\partial}{\partial x})\left(\frac{\partial}{\partial z}\left(\frac{1}{\gamma(\frac{\partial}{\partial x})}\cdot\frac{\partial\gamma(\frac{\partial}{\partial x})}{\partial y_k^i}\right)\right)=\lambda_i\left(tr_{\hat{h}_z}\left(d_{\hat{\nabla}^z}\left(\frac{d\hat{h}_z^i}{dz}\right)\left(.,J^i.,\widehat{\frac{\partial}{\partial y_k^i}}\right)\right)\right)\not\equiv 0$$

   for $i=1,\dots,r$, $k=1,\dots,d_i$ and $\hat{\nabla}^z$ the Levi-Civita connection with respect to $\hat{h}_z$.

4. *For $i=r+1,\dots,s$ and $k=1,\dots,d_i$ holds*

$$\frac{\partial\gamma(\frac{\partial}{\partial x})}{\partial y_k^i}=0 \text{ and } \frac{\partial\gamma(\frac{\partial}{\partial x})}{\partial x}=0.$$

The $\widehat{\ }$ denotes the projection on the Riemannian manifolds. The expression $d_\nabla B$ is defined for a symmetric $(2,0)$–tensor field as follows

$$d_\nabla B(U,V,W):=\nabla_U B(V,W)-\nabla_V B(U,W).$$

The necessary occurrence of an irreducible component of $\hat{\mathfrak{h}}$ which is contained in $\mathfrak{u}(d_i)$ follows from the fact that irreducible subalgebras of $\mathfrak{so}(d_i)$ which are not contained in $\mathfrak{u}(d_i)$ are semisimple and does not have a center (see chapter 4). But this is necessary for the coupled types.
These conditions yield the following

2.20 Corollary. *[Bou00] Let (M,h) be an indecomposable, non-irreducible Lorentzian manifold whose metric is given in the form of theorem 2.17 around a point $p \in M$. If it satisfies the additional condition, that there is an $1 \le i \le s$ with*

1. *The action of $\hat{\mathfrak{h}}$ on $\hat{\Upsilon}^i_p$ is unitary with respect to $\hat{h}^i := h_{|\hat{\Upsilon}^i_p}$ and a complex structure J^i,*
2. *The center of $\mathfrak{u}(\hat{\Upsilon}^i_p, h^i, J^i)$ is contained in $\hat{\mathfrak{h}}_{|\hat{\Upsilon}^i_p}$,*
3. *All the leaves of the foliation $\hat{\mathcal{Y}}^i_q$ equipped with a Riemannian metric by the coordinates (see the previous section) are Ricci-flat for every $q \in U$.*

If the functions $\gamma(\frac{\partial}{\partial y^i_k})$ are given, then $\gamma(\frac{\partial}{\partial x})$ can be chosen in a way that the holonomy of (M,h) is of type 3.
An analogous result is true for type 4.

The difficult point of this corollary is to construct examples which satisfy the conditions 2 and 3 simultaneously.

2.3 Walker coordinates and holonomy

In this section we will describe the holonomy of an indecomposable Lorentzian manifold by means of Walker coordinates. For Walker coordinates defined by a family of Riemannian metrics with conjugated holonomy we will describe the condition that the $SO(n)$-projection of the Lorentzian holonomy is equal to that Riemannian holonomy in terms of the local form of the metric.
Let $p \in M$ arbitrary. We fix Walker coordinates (U,φ) around p, such that $\varphi(p) = 0$ and

$$h = 2dxdz + fdz^2 + \underbrace{\sum_{k=1}^{n} u_k dy_k\, dz}_{=:\phi_z} + \underbrace{\sum_{k,l=1}^{n} g_{kl} dy_k dy_l}_{=:g_z}.$$

With the help of this local description in [Ike96] the following proposition is proved on the holonomy of the Riemannian manifolds

$$\left(W_{(x,z)} = \{\varphi^{-1}(x, y_1, \ldots, y_n, z) | y_i \in \mathbb{R}\} \;,\; g_z = h_{|W_{(x,z)}} \right).$$

2.21 Proposition. *Let $p = \varphi^{-1}(x, y_1, \ldots, y_n, z) \in W_{(x,z)} \subset U$. Then $Hol_p(W_{(x,z)}, g_z) \subset pr_{SO(n)}\left(Hol^0_p(M,h)\right)$.*

It is not clear whether it holds equality, or under which conditions it holds equality.
In the following we will give conditions for an indecomposable, non-irreducible Lorentzian manifold to have holonomy of the type

$$(\mathbb{R} \oplus \text{ Riemannian holonomy}) \ltimes \mathbb{R}^n.$$

2.3.1 The projected connection

The Walker coordinates define Riemannian manifolds $(W_{(x,z)}, g_z)$ as described above. With respect to these we introduce a

2.22 Notation. For an arbitrary point $q \in U$ and an element in the fibre $s := (X, s_1, \ldots, s_n, Z) \in \mathcal{P}_q(M,h)$ we set $\hat{s} := (\hat{s}_1, \ldots, \hat{s}_n) \in \mathcal{O}^{(x(q),z(q))}$ with

$$\hat{s}_i := \text{projection of } s_i \text{ on the tangent space } T_q W_{(x(q),z(q))}.$$

If σ is a local section in $\Gamma(\mathcal{P}(M,h))$ we denote by $\hat{\sigma}$ the (x,z)–dependent family of sections in $\mathcal{O}(W_{(x,z)}, g_z)$ defined by $\hat{\sigma}(q) = \widehat{\sigma(q)}$.

Let $\omega : T\mathcal{P}(M,h) \to \mathfrak{p}$ be the Levi-Civita connection form of (M,h). We will study the projection to the $\mathfrak{so}(n)$–component. First we will prove an obvious property.

2.23 Lemma. *For $\sigma \in \Gamma(\mathcal{P}(M,h))$ and the local connection form ω^σ hold that $pr_{\mathfrak{so}(n)} \circ \omega^\sigma$ does depend only on $\hat{\sigma}$.*

Proof. The local connection form is given by

$$\begin{aligned}
\omega^\sigma(U) &= h(\nabla_U \xi, \zeta)\,(E_{00} - E_{n+1n+1}) && \text{(the } \mathbb{R}\text{–part)} \\
&\quad + \sum_{k=1}^{n} h(\nabla_U \sigma_k, \zeta)\,(E_{0k} - E_{kn+1}) && \text{(the } \mathbb{R}^n\text{–part)} \\
&\quad + \sum_{1 \le k < l \le n} h(\nabla_U \sigma_k, \sigma_l) D_{kl} && \text{(the } \mathfrak{so}(n)\text{–part).}
\end{aligned}$$

for $U \in TM$. Since $\sigma_i \in \Gamma(\Xi^\perp)$ and Ξ as well as $\Xi^\perp$ are parallel it is

$$h(\nabla \sigma_i, \sigma_j) = h(\nabla \sigma_i, \hat{\sigma}_j) = h(\nabla \hat{\sigma}_i, \hat{\sigma}_j),$$

i.e. the the projection

$$pr_{\mathfrak{so}(n)} \circ \omega^\sigma = \sum_{1 \le k < l \le n} h(\nabla_U \sigma_k, \sigma_l) D_{kl} = \sum_{1 \le k < l \le n} h(\nabla_U \hat{\sigma}_k, \hat{\sigma}_l) D_{kl}$$

does only depend on $\hat{\sigma}$ □

We introduce a further

2.24 Notation. Let ω be the Levi-Civita connection of (M,h) and $\sigma \in \Gamma(\mathcal{P}(M,h)_{|U})$ for Walker coordinates $(U, \varphi = (x, y_1, \ldots, y_n, z))$. Then we set

$$\begin{aligned}
\omega^{\hat{\sigma}} &:= pr_{\mathfrak{so}(n)} \circ \omega^\sigma : T\mathcal{P}(M,h) \to \mathfrak{so}(n) \\
\omega^{\hat{\sigma}}_z &:= \omega^{\hat{\sigma}}\left(\frac{\partial}{\partial z}\right) : M \to \mathfrak{so}(n)
\end{aligned}$$

For these objects holds the following

2.25 Lemma. *1. For $\omega^{\hat{\sigma}}$ holds the following transformation formula: If σ and ρ are local sections such that $\sigma = \rho \cdot A$ with $A \in C^\infty(M, (\mathbb{R} \times SO(n)) \ltimes \mathbb{R}^n)$ then*

$$\omega^{\hat{\sigma}}_q = Ad(pr_{\mathfrak{so}(n)}(A_q^{-1})) \circ \omega^{\hat{\rho}_q} + pr_{\mathfrak{so}(n)}\left((dL_{A^{-1}(q)} \circ dA)_q\right).$$

2. Let ϕ_z be the form defined by the Walker coordinates. Then holds

$$\left(\omega^{\hat{\sigma}}_z\right)_{|W_{(x,z)}} = \frac{1}{2} \sum_{i \leq i < j \leq n} \left(d\phi_z(\hat{\sigma}_i, \hat{\sigma}_j) + h\Big([\frac{\partial}{\partial z}, \hat{\sigma}_i], \hat{\sigma}_j\Big) - h\Big([\frac{\partial}{\partial z}, \hat{\sigma}_j], \hat{\sigma}_i\Big)\right) D_{ij}.$$

Proof. The transformation formula follows from the transformation formula for local connection forms and the fact, that Ad in P commutes with $pr_{\mathfrak{so}(n)}$.
The second is a direct calculation using the Koszul formula:

$$\begin{aligned} 2h(\nabla_{\frac{\partial}{\partial z}} \hat{\sigma}_i, \hat{\sigma}_j) &= \hat{\sigma}_i\Big(h(\hat{\sigma}_j, \frac{\partial}{\partial z})\Big) - \hat{\sigma}_j\Big(h(\hat{\sigma}_i, \frac{\partial}{\partial z})\Big) - h([\hat{\sigma}_i, \hat{\sigma}_j], \frac{\partial}{\partial z}) \\ &\quad + h([\frac{\partial}{\partial z}, \hat{\sigma}_i], \hat{\sigma}_j) - h([\frac{\partial}{\partial z}, \hat{\sigma}_j], \hat{\sigma}_i) \\ &= d\phi_z(\hat{\sigma}_i, \hat{\sigma}_j) + h([\frac{\partial}{\partial z}, \hat{\sigma}_i], \hat{\sigma}_j) - h([\frac{\partial}{\partial z}, \hat{\sigma}_j], \hat{\sigma}_i). \end{aligned}$$

□

We will now prove three obvious lemmata verifying the existence of certain local sections under certain conditions.

2.26 Lemma. *There exists a local section $\sigma \in \Gamma(\mathcal{P}(M,h))$ such that $\omega^{\hat{\sigma}}(\frac{\partial}{\partial x}) = 0$.*

Proof. It is $\hat{\sigma}_i = \sum_{k=1}^{n} \eta_{ik} \frac{\partial}{\partial y_k}$ such that $h(\hat{\sigma}_i, \hat{\sigma}_j) = \delta_{ij}$. Since g_{kl} does not depend on x we get

$$0 = \frac{\partial}{\partial x}(h(\hat{\sigma}_i, \hat{\sigma}_j)) = \sum_{k,l=1}^{n} \frac{\partial}{\partial x}(\eta_{ik}\eta_{jl})\, g_{kl}.$$

Thus we can chose η_{ij} with $\frac{\partial}{\partial x}(\eta_{ij}) = 0$. This implies

$$\left[\frac{\partial}{\partial x}, \hat{\sigma}_i\right] = \sum_{k=1}^{n} \frac{\partial}{\partial x}(\eta_{ik}) \frac{\partial}{\partial y_k} = 0.$$

The section $\sigma = (\xi, \sigma_1, \ldots, \sigma_n, \zeta)$ of $\mathcal{P}(M,h)$ then can be defined by

$$\left.\begin{aligned} \xi &:= \tfrac{\partial}{\partial x} \\ \zeta &:= \tfrac{\partial}{\partial z} - \tfrac{f}{2}\tfrac{\partial}{\partial x} \\ \sigma_i &:= \hat{\sigma}_i - \phi_z(\hat{\sigma}_i)\tfrac{\partial}{\partial x}. \end{aligned}\right\} \qquad (2.20)$$

Thus we get the proposition by the Koszul formula and the fact that $\hat{\sigma}_i \in \Gamma(\Xi^\perp)$ and $\Xi^\perp$ integrable.

□

The next lemma illustrates the meaning of the form ϕ_z in the Walker coordinates.

2.27 Lemma. *If the fixed Walker coordinates are such that the forms ϕ_z are closed for every parameter z, then the following holds*

1. *There exists Walker coordinates $\tilde{x}, \tilde{y}_1, \dots \tilde{y}_n, \tilde{z}$ with the properties that $\widetilde{\phi_{\tilde{z}}} = 0$, $\tilde{g}_{ij} = g_{ij}$ and $\tilde{f} = f + \alpha$ with $\alpha \in C^\infty(M)$ and $\frac{\partial}{\partial x}\alpha = 0$.*
2. *There exists a local section $\sigma = (\xi, \sigma_1, \dots, \sigma_n, \zeta) \in \Gamma(\mathcal{P}(M,h))$ with $\omega^{\hat{\sigma}}\left(\frac{\partial}{\partial x}\right) = 0$ such that the distribution $\Sigma = span(\sigma_1, \dots, \sigma_n)$ is integrable.*

The Riemannian manifolds $W_{(\tilde{x},\tilde{z})}$ are equal to leaves of the distribution Σ with the Riemannian metric defined by the orthonormal frame $(\sigma_1, \dots, \sigma_n)$.

Proof. Since ϕ_z are closed — considered as a family of differential forms on $W_{(x,z)}$ — they are a differential of a function φ which does not depend on the x coordinate. More exactly: If $\phi_z = \sum_{k=1}^n u_k dy_k$ with $\frac{\partial}{\partial x}(u_k) = 0$ and

$$0 = d\phi_z = \sum_{l=1}^n du_l \wedge dy_l = \sum_{k,l=1}^n \frac{\partial}{\partial y_k}(u_l) dy_k \wedge dy_l$$

then exists a $\beta \in C^\infty(M)$ with $\frac{\partial}{\partial x}(\beta) = 0$ and $u_k = \frac{\partial}{\partial y_k}(\beta)$.
Now we consider the following coordinates

$$\tilde{x} = x + \beta \ , \ \tilde{y}_i = y_i \ , \ \tilde{z} = z. \tag{2.21}$$

For these holds $\tilde{u}_i = 0$ and $\tilde{f} = f - 2\frac{\partial}{\partial z}(\beta)$, which proves the first point. The second follows by setting

$$\begin{aligned} \xi &:= \frac{\partial}{\partial x} \\ \zeta &:= \frac{\partial}{\partial z} - \frac{f}{2}\frac{\partial}{\partial x} \\ \sigma_i &:= \sum_{k=1}^n \eta_{ik} \frac{\partial}{\partial y_k} \end{aligned}$$

with η_{ik} as in the previous lemma. $\sigma = (\xi, \sigma_1, \dots, \sigma_n, \zeta)$ is a section in $\mathcal{P}(M,h)$. □

2.28 Remark. In general the coordinate transformations which transform Walker coordinates in Walker coordinates are of the form (see [Sch74])

$$\tilde{x} = x + \alpha(z, y_i) \ , \ \tilde{y}_i = \Psi_i(z, y_i) \ , \ \tilde{z} = z. \tag{2.22}$$

Now we make an additional assumption on the metric g_z.

2.29 Lemma. *If the Walker coordinates are such that*

$$g_z = e^{2\gamma} g \tag{2.23}$$

where the metric g does neither depend on x nor on z and γ a function only of the parameter z, then the following holds

1. *There exists a local section $\sigma \in \Gamma(\mathcal{P}(M,h))$ with*

$$\omega^{\hat{\sigma}}\left(\frac{\partial}{\partial x}\right) = 0 \quad \text{and} \quad h\left(\left[\frac{\partial}{\partial z}, \hat{\sigma}_i\right], \hat{\sigma}_j\right) = -\delta_{ij}\ \gamma', \tag{2.24}$$

$$\text{i.e.} \quad \omega^{\hat{\sigma}}_z \;=\; \frac{1}{2}\sum_{1\leq i>j\leq n} d\phi_z(\hat{\sigma}_i,\hat{\sigma}_j) D_{ij} \tag{2.25}$$

2. *There exists Walker coordinates $\tilde{x}, \tilde{y}_1, \dots \tilde{y}_n, \tilde{z}$ with*

$$h = 2d\tilde{x}d\tilde{z} + \tilde{f}d\tilde{z}^2 + \tilde{\phi}_z d\tilde{z} + g,$$

where $\frac{\partial}{\partial x}(\tilde{f}) = \frac{\partial}{\partial x}(f)$ and g are the z–independent terms from (2.23). With the help of these coordinates one constructs a local section $\sigma \in \Gamma(\mathcal{P}(M,h))$ with $\omega^{\hat{\sigma}}\left(\frac{\partial}{\partial x}\right) = 0$ and $[\frac{\partial}{\partial z}, \sigma_i] \in \Gamma(\Xi)$.

Proof. 1.) We define the following section $\sigma = (\xi, \sigma_1, \dots, \sigma_n, \zeta)$ of $\mathcal{P}(M,h)$

$$\left.\begin{array}{rcl} \xi & := & \frac{\partial}{\partial x} \\ \zeta & := & \frac{\partial}{\partial z} - \frac{f}{2}\frac{\partial}{\partial x} \\ \sigma_i & := & \hat{\sigma}_i - \phi_z(\hat{\sigma}_i)\frac{\partial}{\partial x}. \end{array}\right\} \tag{2.26}$$

with $\hat{\sigma}_i = e^{-\gamma}\sum_{k=1}^n \eta_{ik}\frac{\partial}{\partial y_k}$ such that $g(\hat{\sigma}_i, \hat{\sigma}_j) = e^{-2\gamma}\delta_{ij}$ and η_{ik} does neither depend on x nor on z. This is possible since g does neither depend on x nor on z. Thus $\omega^{\hat{\sigma}}\left(\frac{\partial}{\partial x}\right) = 0$. Furthermore it is

$$\left[\frac{\partial}{\partial z}, \hat{\sigma}_i\right] = -\gamma'\ e^{-\gamma}\sum_{k=1}^n \eta_{ik}\frac{\partial}{\partial y_k} = -\gamma\hat{\sigma}_i$$

But this implies $h([\frac{\partial}{\partial z}, \sigma_i], \sigma_j) = -\gamma'\ h(\hat{\sigma}_i, \sigma_j) = -\gamma'\ h(\sigma_i, \sigma_j)$. But this entails (2.25) by lemma 2.25.

2.) We note that the coordinates defined by

$$\tilde{x} = x,\ \tilde{y}_i = \gamma(z)y_i\ ,\ \tilde{z} = z. \tag{2.27}$$

satisfy the properties of the second point. With respect to these coordinates we define a local section σ due to formula (2.26). These satisfy the wanted property. □

2.30 Remark. In general it is not possible to combine the first points of lemma 2.27 and 2.29 to get rid of a closed form ϕ_z and a z–dependent factor simultaneously, because the transformation (2.27) yields a new form $\tilde{\phi}$ which must not necessarily closed. Also by a more general form of (2.27) it cannot be achieved that the old and the transformed ϕ differ by a differential.

Nevertheless we can use these lemmata to study the connection between local coordinates and the holonomy of an indecomposable Lorentzian manifold with recurrent lightlike vector field.

2.3.2 Families of Riemannian metrics

In this section we will describe the condition on the local shape of an indecomposable, non-irreducible Lorentzian manifold to have holonomy of type

$$(\mathbb{R} \times \text{Riemannian holonomy}) \ltimes \mathbb{R}^n$$

in terms of the projected connection. We have to restrict to coordinates which are defined by families of G-metrics.

Locally a z–dependent family of Riemannian metrics is defined by given Walker coordinates. Although the $SO(n)$ projections of the Lorentzian holonomy group at different points p and q in M are conjugated in $O(n)$ this must not be true for the holonomy groups of $g_{z(p)}$ and $g_{z(q)}$. In general we have the situation

$$\begin{array}{ccc} pr_{SO(n)}Hol_p(M,h) & \overset{\text{conjugation}}{\sim} & pr_{SO(n)}Hol_p(M,h) \\ \cup & & \cup \\ Hol_p(W_{(x(p),z(p))}, g_{z(p)}) & \overset{\text{in general}}{\not\simeq} & Hol_q(W_{(x(q),z(q))}, g_{z(q)}) \end{array}$$

Roughly speaking there are two possibilities of changes of the holonomy group of g_z depending on the parameter z:

- The group structure changes by variation of the parameter z.
- The group structure remains the same but the representation of the group changes by variation of the parameter z.

  (a) The family of representation contains only equivalent representations.

  (b) The representations in the family are not equivalent.

2.31 Remark. We do not know whether the case (b) can occur. For example, is it possible that g_{z_1} has the holonomy $U(n)$ given by the representation on the symmetric space $Sp(n,\mathbb{R})/U(n)$, i.e. $U(n) \subset SO(n(n+1))$ and g_{z_2} has holonomy $U(n)$ represented in the standard way, i.e $U(n) \subset SO(2n) \subset 1 \times SO(2n) \subset SO(n(n+1))$, for $n > 2$?

We will not deal with this question but consider the case (a) where we have a z–dependent family of equivalent representations, such that in the above diagram the $\not\simeq$ can be replaced by a $\overset{\text{conjugated}}{\sim}$.

2.32 Definition. Let (U,φ) be Walker coordinates of an indecomposable, non-irreducible Lorentzian manifold (M,h) and let G be a Riemannian holonomy group. We say that the Walker coordinates are defined by a **family of G–metrics** if and only if $Hol_p(W_{(x(p),z(p))},g_{z(p)})$ and $Hol_q(W_{(x(q),z(q))},g_{z(q)})$ are conjugated in $SO(n)$ for every p and q in U.

2.33 Notation. We denote by θ_z the family of Levi-Civita connection forms of g_z on the bundle $\mathcal{O}^{(x,z)} := \mathcal{O}(W_{(x,z)},g_z)$. For a fixed point $\hat{s}_0 \in \mathcal{O}^{(x,z)}_q$ we have the holonomy group $Hol_{\hat{s}_0}(\mathcal{O}^{(x,z)},\theta_z)$. The holonomy bundle of θ_z is defined by

$$\mathcal{H}^{(x,z)}(\hat{s}_0) := \left\{ \hat{s} \in \mathcal{O}^{(x,z)} | \text{There is a } \theta_z\text{–horizontal curve from } \hat{s}_0 \text{ to } \hat{s} \right\}$$

This bundle has structure group $Hol_q(W_{(x,z)},g_z) = Hol_{\hat{s}_0}(\mathcal{O}^{(x,z)},\theta_z)$, which is conjugated to G, and is a reduction of $\left(\mathcal{O}^{(x,z)},\theta_z\right)$.

Then holds the following

2.34 Lemma. *(U,φ) are Walker coordinates around $p \in M$, defined by a family of G–metrics if and only if there exist a locals section $\tau \in \Gamma(\mathcal{P}(M,h))$ such that*

$$Hol_{\tau(q)}(\mathcal{O}(W_{(x,z)},g_z)) = G \tag{2.28}$$

for $q = \varphi^{-1}(x,0,\dots,0,z)$.

Proof. W.l.o.g it is $Hol_s(\theta_0) = G \subset SO(n)$ for $s \in \mathcal{O}_p(W_{(0,0)},g_0))$. The Walker coordinates are defined by a family of G–metrics if and only if for a every local section $\sigma \in \Gamma(\mathcal{P}(M,h))$ with $\sigma(p) = s$ there exists a smooth mapping $A \in C^\infty(\mathbb{R},SO(n))$ such that

$$G = Hol_{\sigma(p)}(\theta_0) = A(z) \cdot Hol_{\sigma(q)}(\theta_z) \cdot A(z)^{-1}$$

for $q = \varphi^{-1}(x,0,\dots,0,z)$ with $A(0) = Id$. For a fixed local section σ we get the section τ of $\Gamma(\mathcal{P}(M,h))$ defined by $\tau(q) = \sigma(q) \cdot A(z(q))$. Hence
$G = A(z) \cdot Hol_{\sigma(q)}(\theta_z) \cdot A(z)^{-1} = A(z) \cdot A(z)^{-1} \cdot Hol_{\tau(q)}(\theta_z) \cdot A(z)^{-1} \cdot A(z) = Hol_{\tau(q)}(\theta_z)$ for $q = \varphi^{-1}(x,0,\dots,0,z)$. □

From now on we fix such a $\tau \in \Gamma(\mathcal{P}(M,h))_{|U}$ with the property (2.28). We say that τ realizes the holonomy G over $q = \varphi^{-1}(x,0\dots,0,z)$. Now we define the following bundle

$$\mathcal{P}^\tau = \dot{\bigcup_{q\in U}} \left\{ s \in \mathcal{P}_q(M,h) \;\middle|\; \begin{array}{l} \hat{s} \in \mathcal{H}^{(x(q),z(q))}_q(\hat{s}_0) \\ \text{with } s_0 = \tau(\varphi^{-1}(x(q),0,\dots,0,z(q)) \end{array} \right\}. \tag{2.29}$$

This is a principal fibre bundle with structure group $(\mathbb{R} \times G) \ltimes \mathbb{R}^n$ since we are in case of families of G-metrics.
Since $(\mathbb{R} \oplus \mathfrak{g}) \ltimes \mathbb{R}^n \subset (\mathbb{R} \oplus \mathfrak{so}(n)) \ltimes \mathbb{R}^n$ is a reductive pair, $pr_{(\mathbb{R}\oplus\mathfrak{g})\ltimes\mathbb{R}^n} \circ w$ restricted to $T\mathcal{P}^\tau$ defines a connection form of this bundle. Furthermore it holds that the connection ω reduces to $(\mathbb{R}^* \times G) \ltimes \mathbb{R}^n$ if and only if the restriction of ω to $\mathcal{P}^\tau$ maps into $(\mathbb{R}\oplus\mathfrak{g}) \ltimes \mathbb{R}^n$, in particular if and only if $pr_{\mathfrak{so}(n)} \circ \omega$ maps into $\mathfrak{g}$.
In the following we will find conditions for the existence of this reduction of the Levi-Civita connection ω of (M, h).
For a vertical vector in $T\mathcal{P}^\tau$ this is given by definition. So we have to verify this inclusion for a vector which is not vertical. But these are images of sections $\sigma \in \Gamma(\mathcal{P}^\tau)$. Hence for every local section σ the local connection form must map into $(\mathbb{R}\oplus \mathfrak{g}) \ltimes \mathbb{R}^n$. But this is true if it holds for one section because of the transformation formula for local connection forms: For local sections σ and ρ such that $\sigma = \rho \cdot A$ with $A \in C^\infty(M, (\mathbb{R} \times G) \ltimes \mathbb{R}^n)$ holds

$$\omega^\sigma_q = Ad(A_q^{-1}) \circ \omega^\rho_q + (dL_{A^{-1}(q)} \circ dA)_q.$$

But this implies that ω^σ maps into the structure group if and only if ω^ρ does.
Hence in the calculations of the local connection forms we will use the section σ defined in lemma 2.26 satisfying $\omega^{\hat{\sigma}}(\frac{\partial}{\partial x}) = 0$ and in addition

$$\hat{\sigma} \in \Gamma\left(\mathcal{H}^{(x(q),z(q))}(\widehat{\tau}(\varphi^{-1}(x(q), 0, \ldots, 0, z(q))))\right). \qquad (2.30)$$

Then we get for $pr_{\mathfrak{so}(n)} \circ \omega^\sigma = \omega^{\hat{\sigma}}$ that

$$\begin{aligned} \omega^{\hat{\sigma}}(\frac{\partial}{\partial x}) &= 0 \\ \omega^{\hat{\sigma}}(\frac{\partial}{\partial y_k}) &= \sum_{1 \le i < j \le n} h(\nabla_{\frac{\partial}{\partial y_k}} \hat{\sigma}_i, \hat{\sigma}_j) D_{ij} \\ &= \sum_{1 \le i < j \le n} g(\nabla^{g_z}_{\frac{\partial}{\partial y_k}} \hat{\sigma}_i, \hat{\sigma}_j) D_{ij} \quad \text{(Koszul-formula)} \\ &= \theta^{\hat{\sigma}}_z(\frac{\partial}{\partial y_k}) \in \mathfrak{g} \quad \text{since } \sigma \in \mathcal{P}^\tau \end{aligned}$$

Thus the term who decides whether ω reduces to $(\mathbb{R} \oplus \mathfrak{g}) \ltimes \mathbb{R}^n$ is $\omega^{\hat{\sigma}}_z = \omega^{\hat{\sigma}}(\frac{\partial}{\partial z})$. This gives the following

2.35 Proposition. *Let (U, φ) be Walker coordinates in an indecomposable, non-irreducible Lorentzian manifold (M, h), defined by a family of G–metrics and let τ be a section in $\mathcal{P}(M, h)_{|U}$ realizing the holonomy G over $q = \varphi^{-1}(x, 0, \ldots, z)$. Then holds*

$$pr_{SO(n)} Hol_p(U, h) = G \quad \textit{if and only if } \omega^{\hat{\sigma}}_z : M \to \mathfrak{g} \textit{ for } \sigma \in \Gamma(\mathcal{P}^\tau).$$

In the following we would like to use the identification of

$$\begin{array}{rcl}\wedge^2\mathbb{R}^n & \simeq & \mathfrak{so}(n)\\ e_i^*\wedge e_j^* & \mapsto & D_{ij}\end{array}$$

to describe this proposition. But as long as $\omega_z^{\hat{\sigma}}$ contains not only $d\phi_z$ terms but also non-linear commutator-terms (see lemma 2.25) every identification of $\omega_z^{\hat{\sigma}}$ with a 2–form would be dependent of the chosen section σ. Thats why in the following section we will consider a restricted situation in order to find manageable sufficient conditions on the manifold to have holonomy of the type ($\mathbb{R}^*\times$ Riemannian holonomy) $\ltimes \mathbb{R}^n$.

2.3.3 Holonomy with Riemannian holonomy as $SO(n)$–component

In this section we will assume that we have Walker coordinates (U,φ) around p, i.e. $\varphi(p)=0$, with a family of Riemannian metrics given by

$$g_z = e^{2\gamma}g$$

with $\frac{\partial}{\partial x}(g_{ij}) = \frac{\partial}{\partial z}(g_{ij}) = 0$ and γ a function only of the parameter z, as in lemma 2.29. Let now $\hat{s}_0 \in \mathcal{O}_p(W_{(0,0)},g)$ with $\hat{s}_i = \sum_{k=1}^n \eta_{ik}\frac{\partial}{\partial y_k}(p)$ with $\eta_{ik}\in\mathbb{R}$ and such that

$$Hol_{\hat{s}_0}(\mathcal{O}(W_{(0,0)},g)) = G.$$

Then $\hat{s}_z := \sum_{k=1}^n \eta_{ik}\frac{\partial}{\partial y_k}(q)$ with $q=\varphi^{-1}(x,0,\dots,0,z)$ is in $\mathcal{O}_q(W_{(0,0)},g)$ and it holds

$$Hol_{\hat{s}_z}(\mathcal{O}(W_{(x,z)},g)) = G.$$

From lemma 2.25 then follows that

$$\omega_z^{\hat{\sigma}} = \frac{1}{2}\sum_{1\le i>j\le n} d\phi_z(\hat{\sigma}_i,\hat{\sigma}_j)D_{ij}.$$

But this implies that $\omega_z^{\hat{\sigma}}$ defines a 2–form by

$$\omega_z := d\phi_z.$$

The holonomy bundle $\mathcal{H}^{(x,z)}(\hat{s}_z)$ of θ — the Levi-Civita connection form of g, no longer depending on z — can be used to establish the correspondence

$$\begin{array}{rcccl}\Omega^2 W_{(x,z)} & \simeq & \mathcal{H}^{(x,z)}(\hat{s}_z)\times_G \wedge^2\mathbb{R}^n & \simeq & \mathcal{H}^{(x,z)}(\hat{s}_z)\times_G \mathfrak{so}(n)\\ \psi & \mapsto & \left[\hat{\sigma}, \sum\limits_{1\le i<j\le n}\psi(\hat{\sigma}_i,\hat{\sigma}_j)e_i^*\wedge e_j^*\right] & \mapsto & \left[\hat{\sigma}, \sum\limits_{1\le i<j\le n}\psi(\hat{\sigma}_i,\hat{\sigma}_j)D_{ij}\right].\end{array}$$

Now we can apply the proposition in that special situation and get that

$$pr_{SO(n)}(Hol_p(U,h)) = G \text{ if and only if } d\phi_z \in \mathcal{H}^{(x,z)}(\hat{s}_z)\times_G \mathfrak{g},$$

where $\mathcal{H}^{(x,z)}(\hat{s}_z)$ is the holonomy bundle of $\mathcal{O}(W_{(x,z)},g)$ in $\hat{s}_z$.
We get for the global setting

2.36 Theorem. *Let $G \subset SO(n)$ a Lie subgroup and $\mathfrak{g}$ its Lie algebra. Let further (M,h) be an $n+2$-dimensional, simply connected Lorentzian manifold with recurrent lightlike vector field and the property that for every $p \in M$ exist Walker coordinates $(U, \varphi = (x, y_1, \dots, y_n, z))$ with $\varphi(p) = 0$ and*

$$h = 2dxdz + f dz^2 + \phi_z dz + e^{2\gamma} \cdot g$$

where the Riemannian metric g on $W_{(x,z)}$ does not depend on z, and γ is a function depending only on the parameter z. Suppose that G is the holonomy of g and let $q \in M$. Then $Hol_q((M,h))$ is conjugated to $G \ltimes \mathbb{R}^n$ $[(\mathbb{R} \times G) \ltimes \mathbb{R}^n]$ if and only if

- $\frac{\partial}{\partial y_k}(f) \neq 0$ *for every k [and $\frac{\partial}{\partial x}(f) \neq 0$],*
- $d\phi_z \in \mathcal{H}^{(x,z)}(\hat{s}_z) \times_G \mathfrak{g}$ *(with the above notations).*

Proof. Under the conditions on ϕ_z we could prove for every neighborhood that $pr_{SO(n)}(Hol_p(U,h)) = G$. The conditions to the function f ensure that $\mathbb{R}^n$ [and $\mathbb{R}$] are contained in the holonomy. So we have for every coordinate neighborhood that $Hol_p(U,h) = G \ltimes \mathbb{R}^n$ [respectively $= (\mathbb{R} \times G) \ltimes \mathbb{R}^n$].
So we have that the dimension of the holonomy is constant (since it is equal in every coordinate neighborhood). Hence the global holonomy is equal to $(\mathbb{R} \times G) \ltimes \mathbb{R}^n$ or $G \ltimes \mathbb{R}^n$ respectively by theorem 1.6. □

This theorem gives a **construction method** for indecomposable Lorentzian manifolds with prescribed holonomy of type

$$(\mathbb{R} \times \text{Riemannian holonomy}) \ltimes \mathbb{R}^n \text{ resp. } (\text{Riemannian holonomy}) \ltimes \mathbb{R}^n :$$

2.37 Corollary (Construction method). *Let (N,g) be a Riemannian manifold, γ a function of $\mathbb{R}$, $f \in C^\infty(N \times \mathbb{R}^2)$ a function, depending on all variables, and ϕ_z a family of 1-forms on N, satisfying the property $d\phi_z \in \mathcal{H}(N,g) \times_{Hol_p(N,g)} \mathfrak{hol}_p(N,g)$. Then the Lorentzian manifold*

$$\left(M = N \times \mathbb{R}^2, h = 2dxdz + f dz^2 + \phi_z dz + e^{2\gamma} \cdot g\right) \tag{2.31}$$

is indecomposable, non-irreducible and has holonomy $(\mathbb{R}^* \times Hol_p(N,g)) \ltimes \mathbb{R}^n$. If $\frac{\partial}{\partial x}(f) = 0$ then it has holonomy $Hol_p(N,g) \ltimes \mathbb{R}^n$.

To reformulate this corollary for Kähler, Ricci-flat Kähler and hyper-Kähler manifolds we need some

2.38 Notation (Kähler manifolds). Let E be a real vector space with a complex structure J. This yields a decomposition of $E^{\mathbb{C}} =: V = V_+ \oplus V_-$ where $V_\pm$ are the eigen spaces to the eigen values i and $-i$ of $J^{\mathbb{C}}$. It holds $\overline{V}_\pm = V_\mp$. Then the vector space of two forms decomposes

$$\wedge^2 V^* = \wedge^{2,0} V^* \oplus \wedge^{1,1} V^* \oplus \wedge^{0,2} V^*$$

where $\wedge^{2,0}V^* := \wedge^2 V_+^*$, $\wedge^{0,2}V^* := \wedge^2 V_-^*$ and $\wedge^{1,1}V^* = V_+^* \wedge V_-^*$. For such a two form one defines the trace $tr\omega = \sum_{k=1}^{n} \omega(e_i, \bar{e}_i)$ for a basis e_i of V_+. We denote by $\wedge_0^{1,1}V^*$ the traceless forms.

Going back to the real space E every 2–form ω decomposes in $\omega = \omega_+ + \omega_-$ with $\omega_+^{\mathbb{C}} \in \wedge^{1,1}V^*$ and $\omega_-^{\mathbb{C}} \in \wedge^{2,0}V^* \oplus \wedge^{0,2}V^*$. I.e. $\omega_\pm(J., J.) = \pm\omega_\pm$. Considering a basis $(e_i : i = 1, \ldots, 2n$ with $e_{2k} = Je_{2k-1})$ one can define the trace $tr_J\omega := \sum_{k=1}^{n} \omega(e_{2k-1}, e_{2k})$. Its complexification is equal to the trace of the complexified form. We denote by

$$\wedge^{1,1}(V^*, J) := \{\omega \in \wedge V^* | \omega(J., J.) = \omega\} \tag{2.32}$$

$$\wedge_0^{1,1}(V^*, J) := \{\omega \in \wedge^{1,1}(V^*, J) | tr_J\omega = 0\} \tag{2.33}$$

$$\wedge^{-}(V^*, J) := \{\omega \in \wedge V^* | \omega(J., J.) = -\omega\} \tag{2.34}$$

Let now (N, g) be a Riemannian manifold of dimension $2n$. A fixed orthonormal basis s in T_pM defines a complex structure J_s on T_pM via $J_s s_{2k-1} = s_{2k}$ and $J_p s_{2k} = -s_{2k-1}$ such that g_p is invariant under J_s.

Let θ be the Levi-Civita connection form and $Hol_s(\mathcal{O}(N, g), \theta) \subset U(n) \subset SO(2n)$. (With $U(n) \subset SO(2n)$ we refer to the usual inclusion given for example in [BK99], and not any conjugate one.) Then J_s extends to a complex structure J (still depending on s) on M — which makes M to a Kähler manifold — and the holonomy bundle $\mathcal{H}(s)$ is equal to the bundle

$$\mathcal{U}(N, g, J) := \{(s_1, \ldots, s_{2n}) \in \mathcal{O}(N, g) | Js_{2k-1} = s_{2k}\}.$$

For a different basis s' in T_pM the holonomy is conjugated in $SO(2n)$ to $U(n)$ and the the resulting complex structure too.

Again we define the forms

$$\Omega^{1,1}(M, J) := \{\omega \in \Omega^2 V^* | \omega(J., J.) = \omega\} \tag{2.35}$$

$$\Omega_0^{1,1}(M, J) := \{\omega \in \Omega^{1,1}(V^*, J) | tr_J\omega = 0\} \tag{2.36}$$

$$\Omega^{-}(M, J) := \{\omega \in \Omega^2 V^* | \omega(J., J.) = -\omega\}. \tag{2.37}$$

These vector bundles are associated to the $U(n)$ bundle $\mathcal{U}(N, g, J)$ with the corresponding vector spaces.

Now we get the following

2.39 Corollary. *Let (M,h) be a Lorentzian manifold, constructed by method (2.31). Then holds*

1. *If (N,g) is Kähler and $d\phi_z \in \Omega^{(1,1)}(N,J)$, then $pr_{SO(n)}Hol(M,h) \subset U(n)$.*

2. *If (N,g) is Kähler and Ricci-flat, and $d\phi_z \in \Omega_0^{(1,1)}(N,J)$, then $pr_{SO(n)}Hol(M,h) \subset SU(n)$.*

3. *If (N,g) is hyper-Kähler with respect to the complex structures I and J, $d\phi_z \in \Omega^{(1,1)}(N,I)$ and $d\phi_z \in \Omega^{(1,1)}(N,J)$, then $pr_{SO(n)}Hol(M,h) \subset Sp(n)$.*

The same can be performed for every other type of Riemannian holonomy.

2.40 Remark (Example). The importance of the condition on the form ϕ illustrates an example given in [Ike96]. One considers the following one-form $\phi = \sum_{k=1}^{5} u_k dy_k$ on $\mathbb{R}^5$ with

$$\begin{aligned} u_1 &= -4y_1y_2 \\ u_2 &= 4y_1y_2 \\ u_3 &= -y_1y_4 - y_2y_4 + y_1y_3 - y_2y_3 + \sqrt{3}(y_4y_5 - y_3y_5) \\ u_4 &= y_1y_4 - y_2y_4 + y_1y_3 + y_2y_3 + \sqrt{3}(y_4y_5 + y_3y_5) \\ u_5 &= 0. \end{aligned}$$

Then one defines the Lorentzian manifold

$$\left(\mathbb{R}^7 = \mathbb{R} \times \mathbb{R}^5 \times \mathbb{R}, h := dxdz + fdz^2 + \phi dz + \sum_{k=1}^{5} dy_k^2\right)$$

where f is a function on $\mathbb{R}^7$ with $\frac{\partial f}{\partial y_i} \neq 0$. Obviously ϕ does not satisfy the assumptions of corollary 2.31. The family of Riemannian metrics is flat and not depending on the parameter z. But the holonomy of this manifold equals to $(\mathbb{R} \oplus \mathfrak{so}(3,\mathbb{R})) \ltimes \mathbb{R}^5$ or if f does not depend on x equal to $\mathfrak{so}(3,\mathbb{R}) \ltimes \mathbb{R}^5$ where $\mathfrak{so}(3,\mathbb{R}) \subset \mathfrak{so}(5,\mathbb{R})$ is the following irreducible representation. The Lie algebra $\mathfrak{sl}(3,\mathbb{R})$ can be decomposed into vector spaces $\mathfrak{sl}(3,\mathbb{R}) = \mathfrak{so}(3,\mathbb{R}) \oplus sym_0(3,\mathbb{R})$, where $sym_0(3,\mathbb{R})$ denote the trace free symmetric matrices. This is a 5–dimensional vector space, invariant and irreducible under the adjoint action of $\mathfrak{so}(3,\mathbb{R})$. Hence it gives the irreducible inclusion $\mathfrak{so}(3,\mathbb{R}) \subset \mathfrak{so}(5,\mathbb{R})$. (The same can be performed for the compact real form $\mathfrak{su}(3)$ of $\mathfrak{sl}(3,\mathbb{C}$ instead of the non compact one $\mathfrak{sl}(3,\mathbb{R})$.)
On the other hand the holonomy of this example does not have a $\mathfrak{so}(5)$–projection which is not a Riemannian holonomy. Because of the definition of the representation it is the holonomy of the symmetric space $Sl(3,\mathbb{R})/SO(3,\mathbb{R})$, resp. $SU(3)/SO(3,\mathbb{R})$. The study of this example was the starting point for the classification approach in chapter 4.

Chapter 3

Parallel spinors on indecomposable Lorentzian manifolds

In this chapter, on one hand we will draw the consequences which arise by the existence of parallel spinors on indecomposable Lorentzian manifolds. As we saw in chapter 1 a parallel spinor induces a parallel timelike or a parallel lightlike vector field. In case the vector field is timelike the manifold is decomposable — a timelike parallel direction splits. This implies that the Dirac current of a parallel spinor field on an indecomposable Lorentzian manifold is lightlike, i.e. on the manifold it exists a parallel vector field. So we are in case of indecomposable, non-irreducible Lorentzian manifolds. First we will draw some basic algebraic consequences. In particular the fact that the existence of a parallel spinor on an indecomposable Lorentzian manifold with lightlike parallel vector field only depends on the $SO(n)$-component of the holonomy group. Further it does not depend on the type of indecomposability, coupled or uncoupled. In addition for a Lorentzian manifold defined by a family of Ricci-flat Kähler metrics we can show that the property of admitting a parallel spinor forces the $SO(n)$-projection of the holonomy to be $SU(n)$. We will draw some further consequences from theorem 2.7.
On the other hand in the second section will rewrite the equation for a parallel spinor field in terms of Walker coordinates. Using this we will give sufficient conditions to the manifold to admit a parallel spinor field in accordance to the previous chapter.

3.1 Parallel spinors and indecomposable holonomy

3.1.1 Notations

At first we will study the consequences of the existence of a parallel spinor to the holonomy of an indecomposable Lorentzian manifold (M, h). The latter shall be simply connected and $n+2$–dimensional throughout the whole section.
As mentioned frequently, the parallel spinor $\varphi \in \Gamma(S)$ defines a parallel lightlike vector field, here denoted by $X := V_\varphi$ and fixed from now on. So we have first of all another reduction of the bundle $\mathcal{P}(M,h)$ to the bundle

$$\mathcal{P}^X(M,h) := \{(X, s_1, \dots, s_n, Z) \in \mathcal{P}(M,h)\}$$

with structure group $Q = O(n) \ltimes \mathbb{R}^n \subset P \subset O(\mathbb{R}^n, \eta)$ (see section 2.1.1).
A spin structure is defined as a reduction of the bundle of oriented *orthonormal* frames, denoted by $\mathcal{O}(M,h)$, with fibres

$$\mathcal{O}_p(M,h) := \left\{ (s_-, s_+, s_1, \dots, s_n) \;\middle|\; \begin{array}{l} \text{an oriented, orthogonal basis in } T_pM \text{ with} \\ h(s_-, s_-) = -1, h(s_+, s_+) = h(s_i, s_i) = 1 \end{array} \right\}$$

and structure group $O(1, n+1)$. Thats why we will work in this chapter with reductions of that bundle. So we define the principal fibre bundle with respect to the fixed parallel lightlike vector field X

$$\mathcal{Q}(M,h) := \{(s_-, s_+, s_1, \dots, s_n) \in \mathcal{O}(M,h) \mid h(X, s_\pm) = \pm 1, h(X, s_i) = 0\}$$

This bundle again has the structure group $Q := O(n) \ltimes \mathbb{R}^n \subset O(1, n+1)$, i.e. with a different, but equivalent, representation on $\mathbb{R}^{n+2}$. Its Lie algebra is the following — with respect to a basis $(e_-, e_+, e_1, \dots, e_n)$:

$$\mathfrak{q} = \left\{ \left(\begin{array}{ccc} 0 & 0^t & v \\ 0 & 0 & v \\ v & -v^t & A \end{array} \right) \;\middle|\; v \in \mathbb{R}^n, A \in SO(n) \right\}. \tag{3.1}$$

3.1.2 Basic algebraic properties

Now we consider the twofold covering $\lambda : Spin(1, n+1) \to SO(1, n+1)$ and its differential λ_*, which is a Lie algebra isomorphism. It identifies $\mathbb{R}^n \subset \mathfrak{q}$ with the following sub-algebra of $\mathfrak{spin}(1, n+1)$

$$\mathbb{R}^n \stackrel{\lambda_*}{\simeq} span(e_i \cdot (e_- + e_+) | 1 \leq i \leq n) \subset \mathfrak{spin}(1, n+1) \tag{3.2}$$

because of (1.8) of the first chapter.

Now we write the spinor module in the form $\Delta_{1,n+1} = \Delta_n \otimes \Delta_{1,1}$. The spin representation κ can be written with respect to this decomposition for an $A \in \mathfrak{spin}(n) \subset \mathfrak{q}$

$$\kappa_{1,n+1}(A)(u \otimes w) = [\kappa_n(A)(u)] \otimes w. \tag{3.3}$$

In accordance with the notation from chapter 1 we define for a subset C of the Clifford algebra $Cl(r,s)$

$$V_C^{r,s} := \{v \in \Delta_{r,s} \,|\kappa_{r,s}(A)v = 0 \text{ for all } A \in C\} \subset \Delta_{r,s}.$$

Then we get the following proposition, which is purely algebraic.

3.1 Proposition. *Let $\mathbb{R}^n$ and $\mathfrak{h}$ be sub-algebras of $\mathfrak{q} \subset \mathfrak{so}(1,n+1)$ and let $\mathfrak{h}$ be indecomposable. Set $k := [\frac{n}{2}]$. Then the following holds*

1. *$V_{\mathbb{R}^n}^{1,n+1} \simeq \Delta_n$, i.e. $dim V_{\mathbb{R}^n}^{1,n+1} = 2^k$.*
2. *Let $\mathfrak{g} = pr_{\mathfrak{so}(n)}\mathfrak{h}$. Then $V_{\mathfrak{h}}^{1,n+1} \simeq V_{\mathfrak{g}}^n$.*

Proof. 1.) We had $\lambda_*^{-1}(\mathbb{R}^n) = span\,(e_i \cdot (e_- + e_+),\ i = 1,\dots,n)$. With the basis of $\Delta_{1,n+1}$ introduced in chapter 1 $(u(\varepsilon_k,\dots,\varepsilon_0)|\varepsilon_i = \pm 1, i = 0,\dots,k)$ we write an arbitrary element $w \in \Delta_{1,n+1}$ as follows

$$w = w_+ \otimes u(1) + w_- \otimes u(-1) \text{ with } w_\pm \in \Delta_n.$$

From (1.9) follows

$$(e_- + e_+) \cdot u\,(\varepsilon_k,\dots,\varepsilon_0) = (\varepsilon_0 - 1)\,u\,(\varepsilon_k,\dots,\varepsilon_1,-\varepsilon_0)\,,$$

which implies

$$e_i \cdot (e_- + e_+) \cdot w = -2e_i \cdot w_- \otimes u(1).$$

But the e_i's are isomorphisms of $\Delta_{1,n+1}$ and of Δ_n too. So it is

$$V_{\mathbb{R}^n}^{1,n+1} = V_{\{e_i\cdot(e_-+e_+)\}}^{1,n+1} = \Delta_n \otimes V_{\{e_-+e_+\}}^{1,1} = \Delta_n \otimes u(1) \simeq \Delta_n.$$

2.) Here we have to use the distinction of indecomposable subalgebras of the parabolic algebra into four types due to Berard-Bergery and Ikemakhen (here theorem 2.10). Evidently only the types with vanishing $\mathbb{R}$–projection are contained in $\mathfrak{q}$, i.e the types 2 and 4. But both types contain — see theorem 2.10 — a non-trivial sub-algebra of $\mathbb{R}^n$. So there is an $1 \le i \le n$, such that $e_i(e_- + e_+) \in \mathfrak{h}$, which implies that

$$V_{\mathfrak{h}}^{1,n+1} \subset V_{\mathfrak{g}}^n \otimes V_{\{e_-+e_+\}}^{1,1} = V_{\mathfrak{g}}^n \otimes \mathbb{C}u(1).$$

Hence for type 2 the conclusion is obvious. For type 4 the proposition holds because every spinor which is annihilated by one element of $\mathbb{R}^n \subset \mathfrak{spin}(1,n+1)$ is annihilated by any of $\mathbb{R}^n$: We have to consider the Lie algebra $\mathfrak{l}$ which consists of elements of the form $(T, \varphi(T))$ with T in the center of $\mathfrak{g}$ and $\varphi(T) \in \mathbb{R}^n$. If we apply such an element to $w \otimes u(1)$ with $w \in \Delta_n$ we get obviously

$$\kappa_{1,n+1}(T, \varphi(T))(w \otimes u(1)) = (\kappa_n(T)w) \otimes u(1)$$

since $u(1)$ is annihilated by $\varphi(T) \in \mathbb{R}^n$. But this implies the second assertion. □

Concerning the existence of parallel spinors on an indecomposable Lorentzian manifold we conclude that it depends only on the the $\mathfrak{so}(n)$–projection of the holonomy algebra, assuming the $\mathbb{R}$–part to be zero. Concerning the four types of the previous section the existence of parallel spinors gives an obvious restriction.

3.2 Corollary. *Let (M, h) be an indecomposable Lorentzian manifold of dimension $n+2$ with parallel spinor field. Then the holonomy of (M, h) is of type 2 or 4. In both cases the dimension of the space of parallel spinors is equal to $dim V^n_{\mathfrak{g}}$ and depends only on the $\mathfrak{so}(n)$–projection of the holonomy group.*

For the next corollary — which gives a first sufficient condition for the existence of parallel spinors — we need a further definition.

3.3 Definition. A spinor field φ on M is called **pure** if the complex dimension of the space $\{Z \in TM^{\mathbb{C}} | Z \cdot \varphi = 0\}$ is maximal, i.e. equal to $[\frac{dim M}{2}]$ (see for example [LM89]).

3.4 Corollary. *Let (M, h) be a simply connected, indecomposable $n+2$–dimensional Lorentzian manifold with abelian holonomy $\mathbb{R}^n$. Then the space of parallel spinors has dimension $2^{[\frac{n}{2}]}$. All these spinors are pure.*

Proof. Let $\sigma = (\sigma_-, \sigma_+, \sigma_1, \ldots, \sigma_n)$ a local section in the bundle $\mathcal{Q}(M, h)$, and lets denote by $\tilde{\sigma}$ its lift in the spin bundle $\tilde{\mathcal{O}}$. Then the parallel spinors resulting from the basis of $V^{1,n+1}_{\mathbb{R}^n}$ namely $\varphi_{(\varepsilon_k,\ldots,\varepsilon_1)} := [\tilde{\sigma}, u\,(\varepsilon_k, \ldots, \varepsilon_1, 1)]$ are pure. A basis of the space $\{Z \in TM^{\mathbb{C}} | Z \cdot \varphi_{(\varepsilon_k,\ldots,\varepsilon_1)} = 0\}$ is given by $\left(\sigma_- + \sigma_+, \sigma_1 - i\varepsilon_1\sigma_2, \ldots, \sigma_{n-1} - i\varepsilon_{[\frac{n}{2}]}\sigma_n\right)$. □

3.5 Corollary. *Let (M, h) be an indecomposable Lorentzian manifold with parallel spinor field φ. Then $V_\varphi \cdot \varphi = 0$.*

Proof. Let φ be the parallel spinor field. Then there is a section $\tilde{\sigma} \in \Gamma(\tilde{\mathcal{O}})$ such that locally $\varphi = [\tilde{\sigma}, w \otimes u(1)]$. $\tilde{\sigma}$ is the lift of $\sigma = (\sigma_-, \sigma_+, \sigma_1, \ldots, \sigma_n) \in \Gamma(\mathcal{Q}_p(M, h))$ and $V_\varphi = X = \sigma_- + \sigma_+$. Then by definition of the Clifford multiplication we have for every $p \in M$

$$X_p \cdot \varphi_p = [s, e_- + e_+] \cdot [\tilde{s}, w \otimes u(1)] = [\tilde{s}, w \otimes (e_- + e_+) \cdot u(1)] = 0,$$

where $s = \sigma(p)$. □

Concerning the Walker coordinates of a Lorentzian manifold, one gets the following corollary for the Riemannian manifolds by proposition 2.21.

3.6 Corollary. *Let (M,h) be a simply connected, indecomposable Lorentzian manifold with parallel spinor. Then all metrics g_z from the family of metrics given by the local form of h admit parallel spinors.*

3.7 Remark.

1. Since the indecomposable, non-irreducible Lorentzian symmetric spaces with solvable transvection group have holonomy $\mathbb{R}^n$ (see [CW70], also [BI93] for the result) this corollary gives a first group of examples of indecomposable Lorentzian manifold with parallel spinors. In section 3.2.3 we give more examples and describe these spaces further.

2. Corollary 3.2 reduces the problem to find sub-algebras $\mathfrak{h} \subset \mathfrak{so}(1,n+1)$ with $V_{\mathfrak{h}}^{1,n+1} \neq 0$ to the problem to find sub-algebras $\mathfrak{g} \subset \mathfrak{so}(n)$ with $V_{\mathfrak{g}}^{n} \neq 0$.

3. To require that an indecomposable Lorentzian manifold admits a parallel spinor restricts the class of possible Riemannian metrics occurring in the Walker coordinate form very strictly. Since we are in the case of simply connected, complete manifolds no Riemannian locally symmetric spaces can occur, because of the following: The existence of a parallel spinor on Riemannian manifolds entails Ricci-flatness, but this forces a Riemannian symmetric space to be flat. Hence the irreducible factors of the Riemannian metrics are $SU(n)$-, $Sp(n)$-, G_2- and $Spin(7)$-manifolds. In particular if the family of Riemannian metrics is a G family, then G has to be a product of these four groups.

We will now cite some results on the problem which groups can occur as $SO(n)$ projection of an indecomposable Lorentzian manifold with parallel spinor field, achieved in dimensions which are relevant for physics.

3.1.3 Results in low dimensions

Until now necessary conditions for the existence of parallel spinors on a Brinkmann-wave determine the holonomy group are only known in low dimensions. Bryant ([Bry99b] resp. [Bry00], see also [FO00]) has proved a result for dimensions $m \leq 11$ about the maximal groups admitting trivial subrepresentations of the spinor representation.

3.8 Proposition. *[Bry99b] The maximal isotropy groups of a spinor with lightlike* *associated vector under the spin representation of $SO(1,n+1)$ are the following*

$$\begin{aligned} n \leq 3 &: 1 \ltimes \mathbb{R}^n \\ n = 4 &: Sp(1) \ltimes \mathbb{R}^4 \\ n = 5 &: (Sp(1) \times 1) \ltimes \mathbb{R}^5 \\ n = 6 &: SU(3) \ltimes \mathbb{R}^6 \\ n = 7 &: G_2 \ltimes \mathbb{R}^7 \\ n = 8 &: Spin(7) \ltimes \mathbb{R}^8 \text{ and } SU(4) \ltimes \mathbb{R}^8 \\ n = 9 &: (Spin(7) \times 1) \ltimes \mathbb{R}^9 \text{ and } (SU(4) \times 1) \ltimes \mathbb{R}^9 \end{aligned}$$

All groups in this proposition are of the form

$$\text{Riemannian holonomy with parallel spinors} \ltimes \mathbb{R}^n,$$

but it is not known whether all subgroups can be obtained in this way.
Bryant also shows in [Bry99c], resp. [Bry00] the following. For an 11-dimensional Brinkmann-wave $(M, h = dt\, dr + f dr^2 + g_z)$ with a family of Riemannian metrics, the condition on h to have holonomy $(Spin(7) \times 1) \ltimes \mathbb{R}^9$ is equivalent to the following fact: There is an 8-dimensional manifold K^8 with a family of $Spin(7)$-metrics g_z depending on the parameter z, i.e. a 1-parameter family of $Spin(7)$-structures, and the the canonical $Spin(7)$-invariant 4-form Φ_z — of course depending on the parameter — obeys the following equation

$$\frac{\partial}{\partial z}\Phi_z = \alpha\, \Phi_z + \Upsilon,$$

with a function α and an anti-self dual (with respect to g_z) 4-form Υ. Bryant called such a family *conformally anti-self dual.* This condition is locally trivial in the following sense: For a given family of $Spin(7)$-metrics on K^8 there is a z-dependent family of diffeomorphism Ψ_z of K^8 such that $\Psi_z^* g_z$ is a conformally anti-self dual family of $Spin(7)$-metrics.
On the other hand, if one starts with a conformally anti-self dual family of $Spin(7)$-metrics g_z. Then the Lorentzian manifold

$$\left(\mathbb{R}^3 \times K^8, h := 2dxdz + dy^2 + f dz^2 + g_z\right)$$

has holonomy $(1 \times Spin(7)) \ltimes \mathbb{R}^9$ if f is a function of $\mathbb{R}^3 \times K^8$ not depending on x.

Another class of metrics was obtained by Figueroa O'Farril who proved in [FO00] the following. Let (M, h) be a Brinkmann-wave with $h = dxdz + g_z$, $(\hat{M}, \hat{g})$ a Riemannian manifold and $\sigma \in C^\infty(M)$ depending only on z. Then for the holonomy group of the product manifold and the warped product metric $\hat{h} = h + \sigma\hat{g}$ holds the following

$$Hol_{(x,y)}(M \times N, \hat{h}) = \left(Hol_x(M,h) \times Hol_y(\hat{M},\hat{g})\right) \ltimes \mathbb{R}^n.$$

So by choosing both factors with parallel spinors — in [FO00] this is done in dimension 11 — one gets a new class of metrics with parallel spinors, but no new examples of holonomy groups admitting trivial sub-representations which cannot be constructed by the method of corollary 2.37.

3.1.4 Holonomy containing $\mathfrak{su}(n)$

In this section we show the following: if an indecomposable Lorentzian metric, defined by a family of $SU(n)$–metrics admits a parallel spinor, then $pr_{SO(n)}(Hol_p(M,h)) = SU(n)$. This we will achieve by algebraic means.

3.9 Lemma. *Let $A \in \mathfrak{so}(2n)$ and $n \neq 2,4$. If there is a $v \in V^{2n}_{\mathfrak{su}(n)}$ such that $A \cdot v = 0$, then $A \in \mathfrak{su}(n)$.*

Proof. We consider $\mathfrak{u}(n) \subset \mathfrak{so}(2n)$ and the corresponding complex structure J on $\mathbb{R}^{2n}$. Then we have a decomposition

$$\mathfrak{so}(2n) \;=\; \mathfrak{u}(n) \oplus \mathfrak{m}$$

where $\mathfrak{m}$ is the complement, orthogonal with respect to the Killing form of $\mathfrak{so}(2n)$. $\mathfrak{m}$ is isomorphic to $\left(\wedge^2\mathbb{C}^n\right)_{\mathbb{R}}$. If $\mathfrak{u}(n)$ is embedded in $\mathfrak{so}(2n)$ as in [BK99], i.e.

$$\mathfrak{u}(n) = span\left(\begin{array}{c} \{D_{2k-1\ 2l-1} + D_{2k\ 2l} : 1 \le k < l \le n\} \\ \cup \{D_{2k-1\ 2l} - D_{2k\ 2l-1} : 1 \le k \le l \le n\} \end{array}\right) \tag{3.4}$$

then

$$\mathfrak{m} = span\left(\begin{array}{c} \{D_{2k-1\ 2l-1} - D_{2k\ 2l} : 1 \le k < l \le n\} \\ \cup \{D_{2k-1\ 2l} + D_{2k\ 2l-1} : 1 \le k < l \le n\} \end{array}\right). \tag{3.5}$$

An $A \in \mathfrak{so}(2n)$ decomposes then in $A = A_+ + A_-$ with

$$\begin{aligned} A_+ &= \frac{1}{2}(A - JAJ) \in \mathfrak{u}(n) \ \text{ i.e. } A_+J = JA_+ \text{ and} \\ A_- &= \frac{1}{2}(A + JAJ) \in \mathfrak{m} \ \text{ i.e. } A_-J = -JA_- \end{aligned}$$

A_+ decomposes further into the sum of $A_0 \in \mathfrak{su}(n)$ and a multiple of $J = \sum_{k=1}^n D_{2k-12k}$. Now we consider the representation of A on

$$V^{2n}_{\mathfrak{su}(n)} = span(u(\underbrace{\varepsilon,\ldots,\varepsilon}_{n \text{ times}}) \ : \ \varepsilon = \pm 1).$$

A_0 acts trivial by definition. It is $\lambda_*^{-1}(J) = \sum\limits_{k=1}^{n} e_{2l-1} \cdot e_{2l}$ thus

$$\kappa_n(J)u\left(\varepsilon_n,\ldots,\varepsilon_1\right) = -\left(\varepsilon_n + \ldots + \varepsilon_1\right)u\left(\varepsilon_n,\ldots,\varepsilon_1\right) \tag{3.6}$$

and therefore

$$\kappa_n(J)u(\varepsilon,\dots,\varepsilon) = -n\varepsilon u(\varepsilon,\dots,\varepsilon).$$

Hence $J : V^{2n}_{\mathfrak{su}(n)} \to V^{2n}_{\mathfrak{su}(n)}$ changes the signs of one of the components of a vector. For the basis of $\mathfrak{m}$ one calculates for its spin representation on $u(\varepsilon,\dots,\varepsilon) =: u(\varepsilon)$:

$$\begin{aligned}(e_{2k-1}\cdot e_{2l-1} - e_{2k}\cdot e_{2l})\,u(\varepsilon) &= \\ &2(-1)^{l-k-1}\varepsilon^{l-k}u(\underbrace{\varepsilon}_{n},\dots,\varepsilon,\underbrace{-\varepsilon}_{l},\dots\varepsilon,\dots,\varepsilon,\underbrace{-\varepsilon}_{k},\varepsilon,\dots,\underbrace{\varepsilon}_{1}),\\ (e_{2k-1}\cdot e_{2l} + e_{2k}\cdot e_{2l-1})\,u(\varepsilon) &= \\ &-2i\;(-1)^{l-k-1}\varepsilon^{l-k-1}u(\varepsilon,\dots,\varepsilon,-\varepsilon,\varepsilon,\dots,\varepsilon,-\varepsilon,\varepsilon,\dots,\varepsilon).\end{aligned}$$

$$\begin{aligned}\text{I.e.}\quad (e_{2k-1}\cdot e_{2l-1} - e_{2k}\cdot e_{2l})\,u(1) &= -2(-1)^{l-k}u(\dots,-1,\dots,-1,\dots)\\ (e_{2k-1}\cdot e_{2l-1} - e_{2k}\cdot e_{2l})\,u(-1) &= -2u(\dots,1,\dots,1,\dots)\\ (e_{2k-1}\cdot e_{2l} - e_{2k}\cdot e_{2l-1})\,u(1) &= 2i(-1)^{l-k}u(\dots,-1,\dots,-1,\dots)\\ (e_{2k-1}\cdot e_{2l} - e_{2k}\cdot e_{2l-1})\,u(-1) &= -2iu(\dots,1,\dots,1,\dots)\end{aligned}$$

If we set $A_- = \sum_{1\le k<l\le n}\left(a_{kl}\left(D_{2k-1\;2l-1} - D_{2k\;2l}\right) + b_{kl}\left(D_{2k-1\;2l} + D_{2k-1\;2l}\right)\right)$ and $v = v_+u(1,\dots,1) + v_-u(-1,\dots,-1)$ then it is

$$\begin{aligned}\frac{1}{2}A_-\cdot v &= v_+\sum_{1\le k<l\le n}(-1)^{l-k}(ib_{kl}-a_{kl})u(\dots,-1,\dots,-1,\dots)\;(\text{``}\dots\text{''}\,\hat{=}1)\\ &\quad - v_-\sum_{1\le k<l\le n}(ib_{kl}+a_{kl})u(\dots,1,\dots,1,\dots)\;(\text{``}\dots\text{''}\,\hat{=}-1).\end{aligned}$$

Hence for $n\neq 2$ holds $A_-\cdot v\notin V^{2n}_{\mathfrak{su}(n)}$ or zero. In that case $A\cdot v = 0$ implies $A_-\cdot v = 0$. But if $n\neq 4$ this entails $a_{kl}\pm ib_{kl} = 0$, i.e. $a_{kl} = b_{kl} = 0$. But this is equivalent to $A\in\mathfrak{u}(n)$, hence $A\in\mathfrak{su}(n)$. □

From this lemma follows directly

3.10 Proposition. *Let $\mathfrak{g}\subset\mathfrak{so}(2n)$ be a subalgebra with $\mathfrak{su}(n)\subset\mathfrak{g}$. If $V^{2n}_{\mathfrak{g}}\neq 0$ then $\mathfrak{g} = \mathfrak{su}(n)$.*

To complete the *proof* in the remaining two cases $n = 2$ and $n = 4$ we refer to the work of Bryant ([Bry99b], see also [FO00] and the section 3.1.3) who listed the maximal isotropy groups of a spinor up to dimension 9. For $n = 2$ its Lie algebra is $\mathfrak{su}(2)$ and for $n = 4$ it is $\mathfrak{su}(4)$ and $\mathfrak{spin}(7)$ but the latter does not contain $\mathfrak{su}(4)$. □

With respect to parallel spinor fields we can conclude the following

3.11 Theorem. *Let (M,h) be an indecomposable Lorentzian manifold of dimension $2n+2$ with parallel spinor field. If $pr_{\mathfrak{so}(2n)}(\mathfrak{hol}_p(M,h))$ contains $\mathfrak{su}(n)$, then*

$$pr_{\mathfrak{so}(2n)}(\mathfrak{hol}_p(M,h)) = \mathfrak{su}(n).$$

In particular: If (M,h) admits a parallel spinor and the family of Riemannian metrics g_z of the Walker coordinates has holonomy $\mathfrak{su}(n)$ at least for one z — in particular if the Walker coordinates are defined by a family of $\mathfrak{su}(n)$–metrics — then $pr_{\mathfrak{so}(2n)}(\mathfrak{hol}_p(M,h)) = \mathfrak{su}(n)$.

A consequence of this theorem is, that the construction method of Boubel (see corollary 2.20) for manifolds with holonomy of type 4 fails if the Lorentzian manifold admits a parallel spinor and the $SO(n)$-part of the holonomy contains $SU(n)$. This is clear since $SU(n)$ has no center, which is necessary for type 4 holonomy.

3.12 Corollary. *An indecomposable Lorentzian manifold with parallel spinor and $\mathfrak{su}(n) \subset pr_{\mathfrak{so}(2n)}(\mathfrak{hol}_p(M,h))$ has uncoupled holonomy of type 2, equal to $Hol_p(M,h) = SU(n) \ltimes \mathbb{R}^{2n}$.*

3.1.5 Further algebraic consequences

Theorem 2.7 gives us further algebraic conditions for the $\mathfrak{so}(n)$–component of an indecomposable Lorentzian manifold with parallel spinor field.
First of all we get a corollary of the propositions 1.20 and 3.1.

3.13 Corollary. *Let (M,h) be a simply connected, indecomposable $n+2$–dimensional Lorentzian manifold with parallel spinor and*

$$\mathbb{R}^n = E_0 \oplus E_1 \oplus \ldots \oplus E_r \quad \text{and} \quad \mathfrak{g} = \mathfrak{g}_1 \oplus \ldots \oplus \mathfrak{g}_r$$

the decomposition into irreducible subrepresentations of the $\mathfrak{so}(n)$–component of the holonomy due to theorem 2.7. Then for each $i = 1,\ldots,r$ the spin representations of $\mathfrak{g}_i \subset \mathfrak{so}(dimE_i)$ has a trivial subrepresentation.

We can prove more detailed claims.

3.14 Proposition. *Let (M,h) be a simply connected, indecomposable $n+2$–dimensional Lorentzian manifold with parallel spinor. Set $\mathfrak{g} := pr_{\mathfrak{so}(n)}(\mathfrak{hol}_p(M,h))$. Then it holds:*

1. *Let $\mathbb{R}^n = E_0 \oplus E_1 \oplus \ldots \oplus E_r$ and $\mathfrak{g} = \mathfrak{g}_1 \oplus \ldots \oplus \mathfrak{g}_r$ the decomposition into irreducible sub-representations due to theorem 2.7. Then for $i = 1\ldots r$ it has to be dim $E_i = 4r_i$ or $\mathfrak{z}_i = 0$ for $\mathfrak{z}_i$ the center of $\mathfrak{g}_i$.*

2. *If $\mathfrak{g}$ is abelian, then it is trivial.*

Proof. 1.) From theorem 2.8 we have for irreducible acting $\mathfrak{g}_i$ that the center of $\mathfrak{g}_i$ is trivial or $\mathfrak{so}(2)$ and $dim\ E_i = 2k_i$. In case of non-trivial $\mathfrak{z}_i$ we have $\mathfrak{z}_i = \mathbb{R}J$ with $J^2 = -Id$. J is of the form

$$J = \begin{pmatrix} j & 0 & 0 & 0 \\ 0 & j & 0 & 0 \\ 0 & 0 & \ddots & 0 \\ 0 & 0 & 0 & j \end{pmatrix} \quad \text{with } j = \begin{pmatrix} 0 & 1 \\ -1 & 0 \end{pmatrix}.$$

Now it is $\lambda_*^{-1}(J) = \sum_{l=1}^{k_i} e_{2l-1} \cdot e_{2l}$ and therefore

$$\kappa_n(J)u\left(\varepsilon_{k_i}, \ldots, \varepsilon_1\right) = -\left(\varepsilon_{k_i} + \ldots + \varepsilon_1\right) u\left(\varepsilon_{k_i}, \ldots, \varepsilon_1\right).$$

The existence of a parallel spinor implies $V^n_{\mathfrak{z}_i} \neq 0$, thus the k_i's must be even.
2.) Let $\mathfrak{g}$ be abelian. Then it decomposes into $\mathfrak{g} = \mathfrak{t}_1 \oplus \ldots \oplus \mathfrak{t}_r$ and $\mathbb{R}^n = E_0 \oplus E_1 \oplus \ldots \oplus E_r$ as in theorem 2.7. Since $\mathfrak{g}$ is abelian the $\mathfrak{t}_i$'s are abelian, but this entails $\mathfrak{t}_i \simeq \mathfrak{so}(2)$ and $E_i = \mathbb{R}^2$ or $\mathfrak{t}_i$ trivial, for $i = 1, \ldots r$. Now the second proposition follows from the first. □

3.2 Conditions in local coordinates and examples

3.2.1 The local situation

Here we will use the special coordinates of the previous chapter to describe the equation of a parallel spinor. Let (M, h) be an indecomposable, simply connected Lorentzian manifold with parallel spinor field φ with $V_\varphi := X$ lightlike and parallel.
For a local section $\sigma = (\sigma_-, \sigma_+, \sigma_1, \ldots, \sigma_n) \in \Gamma(\mathcal{O}(M, h))$ its lift in the spin bundle $\widetilde{\mathcal{O}}$ is denoted by $\tilde{\sigma}$. A spinor field can be written locally as follows

$$\varphi = [\tilde{\sigma}, w] \quad \text{with } w \in C^\infty(M, \Delta_{1,n+1})$$

and its covariant derivative

$$\begin{aligned} \nabla\varphi &= d\varphi - \frac{1}{2}h(\nabla\sigma_-, \sigma_+)\ \sigma_- \cdot \sigma_+ \cdot \varphi \\ &\quad - \frac{1}{2}\sum_{i=1}^{n} \left(h(\nabla\sigma_-, \sigma_i)\ \sigma_- \cdot \sigma_i \cdot \varphi - h(\nabla\sigma_+, \sigma_i)\ \sigma_+ \cdot \sigma_i \cdot \varphi\right) \\ &\quad + \frac{1}{2}\sum_{1 \le i < j \le n} h(\nabla\sigma_i, \sigma_j)\ \sigma_i \cdot \sigma_j \cdot \varphi. \end{aligned}$$

But for a section σ of $\mathcal{Q}(M, h) := \{(s_-, s_+, s_1, \ldots, s_n) \in \mathcal{O}(M, h) \mid h(X, s_\pm) = \pm 1, h(X, s_i) = 0\}$ it is $\sigma_- + \sigma_+ = X$. This implies

$$0 = \nabla\sigma_- + \nabla\sigma_+ \tag{3.7}$$

and $0 = h(\nabla\sigma_-, \sigma_+) + h(\nabla\sigma_+, \sigma_+) = h(\nabla\sigma_-, \sigma_+) + \frac{1}{2}\underbrace{d\left(h(\sigma_+, \sigma_+)\right)}_{=0}$, i.e.

$$0 = h(\nabla\sigma_-, \sigma_+). \tag{3.8}$$

Using this we get

$$\nabla\varphi = d\varphi + \frac{1}{2}\left(\sum_{i=1}^{n} h(\nabla\sigma_-, \sigma_i)\ \sigma_i \cdot (\sigma_- + \sigma_+) \cdot \varphi + \sum_{1\leq i<j\leq n} h(\nabla\sigma_i, \sigma_j)\ \sigma_i \cdot \sigma_j \cdot \varphi\right).$$

If φ is parallel then — by the previous section — it has the following shape

$$\varphi = [\tilde{\tau}, v_0 \otimes u(1)] \quad \text{with } v_0 \in V^n_{\mathfrak{g}},$$

where τ is a section in the holonomy bundle of (M,h) and $\mathfrak{g} := pr_{\mathfrak{so}(n)}\mathfrak{hol}_p(M,h)$. Let $\tilde{Q} = Spin(n) \ltimes \mathbb{R}^n \subset Spin(1, n+1)$ be the twofold covering of $Q = SO(n) \ltimes \mathbb{R}^n$. On a spinor of the form $v \otimes u(1)$ an element $(\tilde{A}, \tilde{a}) \in Spin(n) \ltimes \mathbb{R}^n$ acts as follows

$$(\tilde{A}, \tilde{a}) \cdot (v \otimes u(1)) = \left(\kappa_n(\tilde{A}^{-1})(v)\right) \otimes u(1).$$

Hence the parallel spinor can be written as

$$\varphi = [\tilde{\tau}, v_0 \otimes u(1)] = [\tilde{\sigma}, v \otimes u(1)]$$

where $v \in C^\infty(M, \Delta_n)$ and $\sigma \in \Gamma(\mathcal{Q}(M,h))$. The equation for the parallel spinor becomes — since $(e_- + e_+) \cdot (v \otimes u(1)) = 0$ —

$$0 \;=\; \nabla[\tilde{\sigma}, v \otimes u(1)] \;=\; \left[\tilde{\sigma}, \left(dv + \frac{1}{2}\sum_{1\leq i<j\leq n} h(\nabla\sigma_i, \sigma_j)\ e_i \cdot e_j \cdot v\right) \otimes u(1)\right]. \tag{3.9}$$

Recalling the notation of $\omega^{\hat{\sigma}}$ from the previous chapter we get:

3.15 Proposition. *A spinor field on (M,h) is parallel if and only if it is locally of the form $\varphi = [\tilde{\sigma}, v \otimes u(1)]$ with $v \in C^\infty(M, \Delta_n)$ obeying the differential equation*

$$dv + \frac{1}{2}\omega^{\hat{\sigma}} \cdot v = 0 \tag{3.10}$$

for a section $\sigma = (\sigma_-, \sigma_+, \sigma_1, \ldots, \sigma_n) \in \Gamma(\mathcal{Q}(M,h))$.

Lemma 2.26 of the previous chapter ensures that we can find a section $\sigma \in \mathcal{Q}(M,h)$ with $[X, \sigma_i] = 0$. If we use such a section to write the equation for the parallel spinor (3.10) implies that for a parallel spinor $[\tilde{\sigma}, v \otimes u(1)]$ holds that

$$dv(X) = 0.$$

Fixing Walker coordinates in a neighborhood implies that v does not depend on the coordinate x, and we can understand it as a family of functions from the Riemannian submanifolds $W_{(x,z)}$ into the spinor module Δ_n. Projecting a section $\sigma \in \Gamma(\mathcal{Q}(M,h))$ onto the tangent spaces of $W_{(x,z)}$ yields a family of sections $\hat{\sigma} \in \Gamma(\mathcal{O}(W_{(x,z)}, g_z)$ and thus a family of spinor fields on $W_{(x,z)}$ by

$$\hat{\varphi}_z := [\, \widetilde{\hat{\sigma}}, v_z \,] \tag{3.11}$$

where $\widetilde{\hat{\sigma}}$ is the lift of $\hat{\sigma}$ into the spin bundle of $(W_{(x,z)}, g_z)$.
Choosing $\hat{\sigma}$ as a section of the holonomy bundle of $(W_{(x,z)}, g_z)$ (3.10) gives that $\hat{\varphi}_z$ is a parallel spinor on $W_{(x,z)}$ for every (x,z).
This proposition implies also corollary 3.6 which asserts that all Riemannian metrics of the Walker coordinates have to admit parallel spinors. In the previous section this was deduced from 2.21.
Differentiating equation (3.10) leads to

$$\begin{aligned} 0 &= d\omega^{\hat{\sigma}} \cdot v + \omega^{\hat{\sigma}} \wedge dv \\ &= (d\omega^{\hat{\sigma}} - \frac{1}{2}\omega^{\hat{\sigma}} \wedge \omega^{\hat{\sigma}}) \cdot v \\ &= pr_{\mathfrak{so}(n)}(\Omega^{\sigma}) \cdot v \end{aligned}$$

where Ω is the curvature of ω. The last equation holds because of the Bianchi-identity and since d commutes with the projection onto $\mathfrak{so}(n)$. But this is a fact, which we already know from the previous sections of this chapter.
But we can use equation (3.10) to get sufficient conditions for the existence of parallel spinors. For this we consider again families of G–metrics admitting parallel spinors, in particular to families of Ricci-flat Kähler and hyper-Kähler-metrics.

3.2.2 Sufficient conditions for families of G–metrics admitting parallel spinors

We consider now Walker coordinates around p defined by a family of G–metrics. Let $\mathfrak{g}$ be the Lie algebra of G. We will now reformulate proposition 3.15 in terms of the space $V^n_{\mathfrak{g}} \subset \Delta_n$.
Analogously to the notation 2.22 of the previous chapter we assign to $s = (s_-, s_+, s_1, \ldots, s_n) \in \mathcal{Q}_q(M,h)$ the element $\hat{s}$ in $\mathcal{O}_q(W_{(x(q),z(q))}, g_{z(q)})$ defined by

$$\hat{s}_i := \text{projection of } s_i \text{ onto the tangent space } T_q W_{(x(q),z(q))}.$$

For a section $\tau \in \Gamma(\mathcal{Q}(M,h))$ we set

$$\mathcal{Q}^{\tau} = \dot{\bigcup_{q \in U}} \left\{ s \in \mathcal{Q}_q(M,h) \;\middle|\; \begin{array}{l} \hat{s} \in \mathcal{H}_q^{(x(q),z(q))}(\hat{s}_0) \\ \text{with } s_0 = \tau(\varphi^{-1}(x(q),0,\ldots,0,z(q)) \end{array} \right\}.$$

Recalling the definition of $\omega_z^{\hat{\sigma}}$ from the previous chapter we get

3.16 Proposition. *Let (U,φ) be Walker coordinates in (M,h) defined by a family of G–metrics which admits parallel spinors on the Riemannian submanifolds $W_{(x,z)}$.*
If there is a local section σ of the bundle $\mathcal{Q}^\tau$, such that the function $\omega_z^{\hat{\sigma}}$ is constant along every $W_{(x,z)}$, then (U,h) admits a parallel spinor.

Proof. Let (U,φ) be Walker coordinates around $p \in M$, defined by a family of G–metrics. Then we can fix a section $\tau \in \Gamma(U, \mathcal{Q}(M,h))$ such that

$$Hol_{\tau(q)}(\mathcal{O}(W_{(x,z)}, g_z)) = G$$

for $q = \varphi^{-1}(x, 0, \ldots, 0, z))$. Then we have a family of parallel spinors over $W_{(x,z)}$. Thus $V_{\mathfrak{g}}^n \neq 0$. Take a section σ of $\mathcal{Q}^\tau$ such that $\omega_z^{\hat{\sigma}}$ is constant on the submanifolds $W_{(x,z)}$ and of course independent of x. Hence $\omega_z^{\hat{\sigma}}$ is in $C^\infty(\mathbb{R})$. Thus we can consider the system of ordinary differential equations for $v \in C^\infty(\mathbb{R}, V_{\mathfrak{g}}^n)$:

$$\frac{dv}{dz} + \omega_z^{\hat{\sigma}} \cdot v(z) = 0. \tag{3.12}$$

This one has a nontrivial solution if we start with an initial value different from zero. A solution $v^\sigma \in C^\infty(\mathbb{R}, V_{\mathfrak{g}}^n)$ then defines a parallel spinor on U by

$$\varphi := [\tilde{\sigma}, v^\sigma \otimes u(1)] .$$

By definition it satisfies equation (3.10) for $\frac{\partial}{\partial z}$ but since $\sigma \in \Gamma(\mathcal{Q}^\tau)$ also for $U \in TW_{(x,z)}$. Hence it is parallel. □

For Walker coordinates, defined by families of Riemannian metrics of the form

$$g_z = e^{2\gamma} g$$

with g independent of z we can weaken the corollary 2.39 a little bit.

3.17 Corollary. *Let (M,h) be a constructed by the method of (2.31) with $g_z = e^{2\gamma} g$ where g does not depend on z.*

1. *If g is a Ricci-flat Kähler metric, $\phi_z \in \Omega^{1,1}(N,J)$ and $tr_J\,(d\phi_z)$ is a function only of z. Then (M,h) admits parallel spinors.*

2. *If g is a hyper-Kähler metric with complex structures $I, J, K = IJ$ and $d\phi_z(X,Y) = d\phi_z(IX,IY) = d\phi_z(JX,JY)$. Then $Hol_x(M,h) \subset Sp(n) \ltimes \mathbb{R}^{4n}$ and (M,h) admits parallel spinors.*

3.2.3 Brinkmann-waves with abelian holonomy

An important example of simply connected, complete, indecomposable Lorentzian manifolds with parallel spinors are manifolds with abelian holonomy $\mathbb{R}^n$. They admit of course a lightlike parallel vector field associated to the parallel spinor, i.e. they are Brinkmann-waves.

A first class of Lorentzian manifolds with abelian holonomy is constituted by the indecomposable Lorentzian symmetric spaces with solvable transvection group (see [CW70], also [BI93] and [Neu02]). The local form of its metric is given by

$$h = 2dxdz + \Big(\sum_{k=1}^{n} a_k y_k^2\Big) dz^2 + \sum_{k=1}^{n} dy_k^2 \quad \text{with } a_k \in \mathbb{R} \text{ constants.}$$

The so-called *pp*-manifolds we study in this section are generalization of indecomposable Lorentzian symmetric spaces with solvable transvection group. They are also generalizations of the plane waves (see [Sch74]).

3.18 Definition. A Brinkmann-wave is called ***pp*-manifold** if for its curvature tensor $\mathcal{R}$ holds the following trace condition

$$tr_{(3,5)(4,6)}(\mathcal{R} \otimes \mathcal{R}) = 0. \tag{3.13}$$

Schimming [Sch74] proved that this condition is equivalent to the existence of local coordinates such that the metric h has the following form

$$h = 2\, dxdz + f dz^2 + \sum_{k=1}^{n} dy_k^2 \quad \text{with } \frac{\partial}{\partial x}(f) = 0.$$

Now we will give necessary and equivalent conditions for a Brinkmann wave to have abelian holonomy.

3.19 Theorem. *A simply connected Brinkmann wave has abelian holonomy $\mathbb{R}^n$ if and only if it is a pp-manifold.*

Proof. Lets assume that a Brinkmann wave (M, h) has holonomy $\mathbb{R}^n$. Let X be the lightlike parallel vector field. We fix $p \in M$ and $s = (X, s_1, \dots s_n, Z) \in \mathcal{P}_p((M, h))$. The holonomy algebra then maps s_i into $\Xi_p = \mathbb{R}\, X$.

So we can check the trace condition with s for $Y_1, \ldots, Y_4 \in T_pM$

$$\begin{aligned}
&tr_{(3,5)(4,6)}(\mathcal{R} \otimes \mathcal{R})_p(Y_1, Y_2, Y_3, Y_4) = \\
&= tr\left(h_p\left(\mathcal{R}_p(Y_1, Y_2)\ .\ , \mathcal{R}_p(Y_3, Y_4)\ .\ \right)\right) \\
&= h_p\Big(\underbrace{\overbrace{\mathcal{R}_p(Y_1, Y_2)}^{\in \mathfrak{hol}_p} X, \overbrace{\mathcal{R}_p(Y_3, Y_4)}^{\in \mathfrak{hol}_p} Z}_{=0}\Big) + h_p\Big(\underbrace{\overbrace{\mathcal{R}_p(Y_3, Y_4)}^{\in \mathfrak{hol}_p} X, \overbrace{\mathcal{R}_p(Y_1, Y_2)}^{\in \mathfrak{hol}_p} Z}_{=0}\Big) \\
&\quad + \sum_{k=1}^{n} h\big(\underbrace{\overbrace{\mathcal{R}(Y_1, Y_2)}^{\in \mathfrak{hol}_p} s_k}_{\in \Xi}, \underbrace{\overbrace{\mathcal{R}(Y_3, Y_4)}^{\in \mathfrak{hol}_p} s_k}_{\in \Xi}\big) \\
&= 0.
\end{aligned}$$

This gives one direction.
The other direction follows from the local form of the metric of a *pp*-manifold and theorem 2.36. □

3.2.4 The Ricci-isotropy

For the Ricci-tensor one can prove the following. The proof is due to [FO00].

3.20 Lemma. *Let (M, h) be a Brinkmann-wave given in Walker coordinates such that*

$$h = 2dxdz + fdz^2 + \phi_z dz + \sum_{k,l=1}^{n} g_{kl} dy_k dy_l.$$

Its Ricci-endomorphism is totally isotropic if and only if $Ric(\frac{\partial}{\partial y_k}, .) = 0$, i.e. $Ric = r\ dz \circ dz$ for a function r.

Proof. By the use of the coordinates we define a section of $\mathcal{P}(M, h)$ (see chapter 2, in particular formula (2.20)).

$$\begin{aligned}
\xi &:= \frac{\partial}{\partial x} \\
\zeta &:= \frac{\partial}{\partial z} - \frac{f}{2}\frac{\partial}{\partial x} \\
\sigma_i &:= \hat{\sigma}_i - \phi_z(\hat{\sigma}_i)\frac{\partial}{\partial x}.
\end{aligned}$$

with $\hat{\sigma}_i = \sum_{k=1}^{n} \eta_{ik} \frac{\partial}{\partial y_k}$ such that $h(\hat{\sigma}_i, \hat{\sigma}_j) = \delta_{ij}$. Then we have

$$\begin{aligned}
h(Ric(U), Ric(V)) &= Ric(U, \xi) Ric(\zeta, V) + Ric(V, \xi) Ric(\zeta, U) \\
&\quad + \sum_{k=1}^{n} Ric(U, \sigma_k) Ric(\sigma_k, V) \\
&= \sum_{k=1}^{n} Ric(U, \hat{\sigma}_k) Ric(\hat{\sigma}_k, V)
\end{aligned}$$

Since the last term is symmetric in U and V we have that $h(Ric(U), Ric(V)) = 0$ if and only if $\sum_{k=1}^{n} Ric(U, \hat{\sigma}_k) Ric(\hat{\sigma}_k, V) = 0$. The latter holds — by the definition of $\hat{\sigma}_i$ — if and only if $Ric(., \frac{\partial}{\partial y_k}) = 0$. This was the claim to prove. □

Supposing the existence of more restricted coordinates for a Brinkmann wave, the isotropy condition on the Ricci endomorphism leads to an interesting equation.

3.21 Proposition. *Let (M, h) be a Brinkmann-wave given in Walker coordinates with flat Riemannian metrics, i.e such that*

$$h = 2dxdz + f dz^2 + \phi\ dz + \sum_{k=1}^{n} dy_k^2.$$

Its Ricci-endomorphism is totally isotropic if

$$d^* d\ \phi_z = 0. \tag{3.14}$$

for all z.

Proof. We consider ϕ_z as a family of forms on the flat $\mathbb{R}^n$. Because of the shape of h, in particular since $g \equiv \delta_{ij}$ we have that $\hat{\sigma}_i = \frac{\partial}{\partial y_i}$ in the above section σ. The lemma and the Koszul formula then give

$$\begin{aligned}
0 &= Ric(\frac{\partial}{\partial z}, \frac{\partial}{\partial y_i}) \\
&= \overbrace{\mathcal{R}((\frac{\partial}{\partial z}, \xi, \zeta, \frac{\partial}{\partial y_i})}^{=0} + \overbrace{\mathcal{R}((\frac{\partial}{\partial z}, \xi, \zeta, \frac{\partial}{\partial y_i})}^{=0} + \sum_{k=1}^{n} \mathcal{R}\left(\frac{\partial}{\partial z}, \frac{\partial}{\partial y_k}, \frac{\partial}{\partial y_k}, \frac{\partial}{\partial y_i}\right) \\
&= -\frac{1}{4}\left[\frac{\partial}{\partial y_k}\left(\frac{\partial}{\partial y_k}(\phi_z(\frac{\partial}{\partial y_i}))\right) - \frac{\partial}{\partial y_k}\left(\frac{\partial}{\partial y_i}(\phi_z(\frac{\partial}{\partial y_k}))\right)\right] \\
&\quad -\frac{1}{2}\left[h\left([\frac{\partial}{\partial y_k}, \nabla_{\frac{\partial}{\partial z}} \frac{\partial}{\partial y_k}], \frac{\partial}{\partial y_i}\right) + h\left([\frac{\partial}{\partial y_i}, \nabla_{\frac{\partial}{\partial z}} \frac{\partial}{\partial y_k}], \frac{\partial}{\partial y_k}\right)\right] \\
&= -\frac{1}{2}\left[\frac{\partial}{\partial y_k}\left(\frac{\partial}{\partial y_k}(\phi_z(\frac{\partial}{\partial y_i}))\right) - \frac{\partial}{\partial y_k}\left(\frac{\partial}{\partial y_i}(\phi_z(\frac{\partial}{\partial y_k}))\right)\right] \\
&= d^* d\phi_z(\frac{\partial}{\partial y_i}).
\end{aligned}$$

Here d^* is the codifferential with respect to the flat Riemannian metric $g \equiv \delta_{ij}$.

□

Chapter 4

Weak-Berger algebras and Lorentzian holonomy

The previous chapters did not answer the question, which representations of which algebras can occur as $\mathfrak{g} := pr_{\mathfrak{so}(n)}(\mathfrak{hol}_p(M,h))$. We have seen that all holonomy algebras of Riemannian manifolds can be realized as $\mathfrak{g}$. The interesting question is, whether these are all. $\mathfrak{g}$ defines a geometric structure on the vector bundle $E := \Xi^{\perp}/\Xi$ (see section 2.1.2, but this structure cannot be understood as a structure of a linear bundle of a manifold. Although the fixing of coordinates defines a family of n–dimensional Riemannian submanifolds, it is not clear how $\mathfrak{g}$ defines a structure on these. The dependence of the Riemannian metrics on z is the point which makes it difficult to classify the possible $\mathfrak{g}$.

This difficulty suggests to go an algebraic way instead of a differential geometric one, in order to classify the possible algebras $\mathfrak{g}$, since $\mathfrak{g}$ depends on the chosen vector Z only by conjugation in $\mathfrak{so}(n)$. We will try to go this algebraic way using Bianchi's first identity, restricted to the representations space of $\mathfrak{g}$. This is the aim of the next sections.

That a further algebraic investigation is possible is a goal of theorem 2.7 of Berard-Bergery and Ikemakhen which has two important consequences.

The first is that it suffices to study irreducible acting groups or algebras $\mathfrak{g}$, a fact which is necessary for trying a classification. We will see this in detail in the first section.

The second is that these irreducible acting, connected subgroups of $SO(n)$ are closed and therefore compact. Hence $G := pr_{SO(n)}Hol_p^0(M,h)$ is compact, although the whole holonomy group must not be compact (for such examples see [BI93]).

On the other hand the relation between the $\mathfrak{so}(n)$–part and the $\mathbb{R}$– and $\mathbb{R}^n$–parts is understood quite well ([Bou00], or very recently [Gal03]): If one has a simply-connected, indecomposable, non-irreducible Lorentzian manifold with holonomy of uncoupled type, then, under certain conditions, one can construct a Lorentzian manifold with coupled

type holonomy. Thus the classification of the $\mathfrak{so}(n)$–component of an indecomposable Lorentzian holonomy group would lead to a classification of holonomy groups of indecomposable, non-irreducible Lorentzian manifolds.
In the first section we formulate an algebraic criterion, which is satisfied by the $\mathfrak{so}(n)$–projection of an indecomposable, non–irreducible Lorentzian holonomy. Because of its analogy to the usual Berger criterion and because it is satisfied by Berger algebras we call Lie algebras obeying this criterion weak-Berger algebras. This property remains under complexification of the representation and the Lie algebra. This we use in the following sections to classify *simple* weak-Berger algebras where the complexified representation remains irreducible and for weak-Berger algebras for which this is not the case.

4.1 Berger algebras, weak-Berger algebras and indecomposable Lorentzian holonomy

4.1.1 Berger and weak-Berger algebras

Here we will introduce the notion of weak-Berger algebras in comparison to Berger algebras. We derive some basic properties, in particular a decomposition property and the behavior under complexification.
Let E be a vector space over the field $\mathbb{K}$ and let $\mathfrak{g} \subset \mathfrak{gl}(E)$ be a Lie algebra. Then one defines:

$$\begin{aligned}\mathcal{K}(\mathfrak{g}) &:= \{R \in \Lambda^2 E^* \otimes \mathfrak{g} \mid R(x,y)z + R(y,z)x + R(z,x)y = 0\} \\ \underline{\mathfrak{g}} &:= span\{R(x,y) \mid x, y \in E, R \in \mathcal{K}(\mathfrak{g})\},\end{aligned}$$

and for $\mathfrak{g} \subset \mathfrak{so}(E,h)$:

$$\begin{aligned}\mathcal{B}_h(\mathfrak{g}) &:= \{Q \in E^* \otimes \mathfrak{g} \mid h(Q(x)y,z) + h(Q(y)z,x) + h(Q(z)x,y) = 0\} \\ \mathfrak{g}_h &:= span\{Q(x) \mid x \in E, Q \in \mathcal{B}_h(\mathfrak{g})\}.\end{aligned}$$

Then we have the following basic properties.

4.1 Proposition. *$\mathcal{K}(\mathfrak{g}) \subset \Lambda^2 E^* \otimes \mathfrak{g}$ and $\mathcal{B}_h(\mathfrak{g}) \subset E^* \otimes \mathfrak{g}$ are $\mathfrak{g}$-modules. $\underline{\mathfrak{g}}$ and $\mathfrak{g}_h$ are ideals in $\mathfrak{g}$.*

Proof. The representation of $\mathfrak{g}$ on $\mathcal{B}_h(\mathfrak{g})$ and $\mathcal{K}(\mathfrak{g})$ is given by the standard and the adjoint representation

$$\begin{aligned}(A \cdot Q)(x) &= -Q(Ax) + [A, Q(x)] \\ (A \cdot R)(x,y) &= -R(Ax,y) - R(x,Ay) + [A, R(x,y)].\end{aligned}$$

Both are modules, because of

$$\begin{aligned}
&h(A\cdot Q(x)y,z)+h(A\cdot Q(y)z,x)+h(A\cdot Q(z)x,y)=\\
&-h(Q(Ax)y,z)-h(Q(Ay)z,x)-h(Q(Az)x,y)\\
&\qquad\qquad +h(AQ(x)y,z)+h(AQ(y)z,x)+h(AQ(z)x,y)\\
&\qquad\qquad\qquad\qquad -h(Q(x)Ay,z)-h(Q(y)Az,x)-h(Q(z)Ax,y) \;=\\
&-h(Q(Ax)y,z)-h(Q(y)z,Ax)-h(Q(z)Ax,y)\\
&\qquad\qquad -h(Q(Ay)z,x)-h(Q(z)x,Ay)-h(Q(x)Ay,z)\\
&\qquad\qquad\qquad\qquad -h(Q(Az)x,y)-h(Q(x)y,Az)-h(Q(y)Az,x) \;=\; 0.
\end{aligned}$$

and for $\mathcal{K}(\mathfrak{g})$ analogously.
This gives that $\underline{\mathfrak{g}}$ and $\mathfrak{g}_h$ are ideals because of

$$\begin{aligned}
{[Q(x),A]} &= (A\cdot Q)(x)+Q(Ax)\in\mathfrak{g}_h \;\text{ for } Q(x)\in\mathfrak{g}_h\\
{[R(x,y),A]} &= (A\cdot R)(x,y)+R(Ax,y)+R(x,Ay)\in\underline{\mathfrak{g}} \;\text{ for } R(x,y)\in\underline{\mathfrak{g}}.
\end{aligned}$$

□

4.2 Definition. Let $\mathfrak{g}\subset\mathfrak{gl}(E)$ be a Lie algebra. Then $\mathfrak{g}$ is is called **Berger algebra** if $\underline{\mathfrak{g}}=\mathfrak{g}$. If $\mathfrak{g}\subset\mathfrak{so}(E,h)$ is an orthogonal Lie algebra with $\mathfrak{g}_h=\mathfrak{g}$, then we call it **weak-Berger algebra**.

Equivalent to the (weak-)Berger property is the fact that there is no ideal $\mathfrak{h}$ in $\mathfrak{g}$ such that $\mathcal{K}(\mathfrak{h})=\mathcal{K}(\mathfrak{g})$ (resp. $\mathcal{B}_h(\mathfrak{h})=\mathcal{B}_H(\mathfrak{g})$).
The notion "weak-Berger" is satisfied by the following

4.3 Proposition. *Every Berger algebra which is orthogonal is a weak-Berger algebra.*

Proof. We have to show the inclusion $\underline{\mathfrak{g}}\subset\mathfrak{g}_h$. If $R(x,y)\in\underline{\mathfrak{g}}$, then a standard calculation gives

$$h(R(x,y)u,v)=h(R(u,v)x,y)=-h(R(u,v)y,x).$$

This means that $R(x,.)\in\mathcal{B}_h(\mathfrak{g})$ for every $x\in E$ and $R\in\mathcal{K}(\mathfrak{g})$. So $R(x,y)\in\mathfrak{g}_h$. □

The proof gives a corollary which we will need later on.

4.4 Corollary. *Let $\mathfrak{g}\subset\mathfrak{so}(E,h)$ be an orthogonal Lie algebra. Then*

$$span\{R(x,y)+Q(z)|R\in\mathcal{K}(\mathfrak{g}),Q\in\mathcal{B}_h(\mathfrak{g}),x,y,z\in E\}\subset\mathfrak{g}_h. \tag{4.1}$$

Proof. $\underline{\mathfrak{g}}\subset\mathfrak{g}_h$ gives the proposition. □

Now we will prove another proposition, which is about the decomposition of Berger and weak-Berger algebras.

4.5 Proposition. *If* $\mathfrak{g}_1 \subset \mathfrak{gl}(V_1)$, $\mathfrak{g}_2 \subset \mathfrak{gl}(V_2)$ *and* $\mathfrak{g} := \mathfrak{g}_1 \oplus \mathfrak{g}_2 \subset \mathfrak{gl}(V := V_1 \oplus V_2)$, *then it holds*

1. *If* $\mathfrak{g}_1$ *and* $\mathfrak{g}_2$ *are Berger algebras, then* $\mathfrak{g}$ *is a Berger algebra.*

2. *If in addition* $\mathfrak{g}_1 \subset \mathfrak{so}(V_1, h_1)$, $\mathfrak{g}_2 \subset \mathfrak{so}(V_2, h_2)$ *and* $\mathfrak{g} := \mathfrak{g}_1 \oplus \mathfrak{g}_2 \subset \mathfrak{so}(V := V_1 \oplus V_2, h := h_1 \oplus h_2)$, *then it holds:*

   $\mathfrak{g}_1$ *and* $\mathfrak{g}_2$ *are weak-Berger algebras if and only if* $\mathfrak{g}$ *is a weak-Berger algebra.*

Proof. The proof for obtaining the property for the sum of two Lie algebras is analogous in both cases. It follows from the obvious inclusions

$$\begin{aligned} \mathcal{B}_{h_1}(\mathfrak{g}_1) \oplus \mathcal{B}_{h_2}(\mathfrak{g}_2) &\subset \mathcal{B}_h(\mathfrak{g}) \subset (V_1 \oplus V_2)^* \otimes (\mathfrak{g}_1 \oplus \mathfrak{g}_2) \quad \text{and} \\ \mathcal{K}(\mathfrak{g}_1) \oplus \mathcal{K}(\mathfrak{g}_2) &\subset \mathcal{K}(\mathfrak{g}). \end{aligned}$$

To show the other direction in the weak-Berger case we consider an $Q \in \mathcal{B}_h(\mathfrak{g})$. Then for $u_i, v_i, w_i \in V_i$, $i = 1, 2$ holds:

$$h(Q(u_1)v_2, w_2) = -h(Q(v_2)w_2, u_1) - h(Q(w_2)u_1, v_2) = 0,$$

since $V_1 \oplus^{\perp} V_2$. So we get $Q(u_1)v_2 = Q(u_2)v_1 = 0$. But this means that $\mathcal{B}_h(\mathfrak{g}) \subset V_1^* \otimes \mathfrak{g}_1 \oplus V_2^* \otimes \mathfrak{g}_2$ and in particular $\mathcal{B}_h(\mathfrak{g}) \subset \mathcal{B}_{h_1}(\mathfrak{g}_1) \oplus \mathcal{B}_{h_2}(\mathfrak{g}_2)$. This implies that $\mathfrak{g}_1$ and $\mathfrak{g}_2$ are weak-Berger if $\mathfrak{g}$ is weak-Berger. □

The Ambrose-Singer holonomy theorem implies that holonomy algebras of torsion free connections — in particular of a Levi-Civita-connection — are Berger algebras. The list of all irreducible Berger algebras is known ([Ber55] for orthogonal, non-symmetric Berger algebras, [Ber57] for orthogonal symmetric ones, and [MS99] in the general affine case).

We should mention that in our notation Berger algebra are not only non-symmetric Berger algebra, as it is sometimes defined. For us only the possibility of being the holonomy algebra is of interest, symmetric or non symmetric.

The $\mathfrak{so}(n)$-projection of an indecomposable, non-irreducible Lorentzian manifold is no holonomy algebra, and therefore not necessarily a Berger algebra. But in the next section we will show that it is a weak-Berger algebra.

4.1.2 Lorentzian holonomy and weak-Berger algebras

Here we prove the following statement.

4.6 Theorem. *Let (M,h) be an indecomposable, but non-irreducible Lorentzian manifold and $\mathfrak{g} = pr_{\mathfrak{so}(n)}(\mathfrak{hol}_p(M,h)) \subset E^* \otimes E$. Then $\mathfrak{g}$ is a weak-Berger algebra.*

Proof. Let $p \in M$ and $\Xi_p \subset \Xi_p^\perp$ fibres of the parallel distributions on M, and let $(X, E_1, \ldots E_n, Z)$ be in $\mathcal{P}_p(M,h)$. We can write now every vector $U \in T_pM$ in this basis as $U = \xi \cdot X + Y + \zeta \cdot Z$ with $Y \in E := span(E_1, \ldots, E_n)$, $\xi, \zeta \in \mathbb{R}$. $(E, h_{p|E})$ is an euclidian vector space.

By the Ambrose-Singer holonomy theorem [AS53] $\mathfrak{hol}_p(M,h)$ is generated by elements of the form

$$A = \mathfrak{p}^{-1} \circ \mathcal{R}(\mathfrak{p}(U), \mathfrak{p}(V)) \circ \mathfrak{p} := \mathcal{R}^{\mathfrak{p}}(U,V) \text{ , for } \mathfrak{p} \text{ the parallel displacement along a piecewise smooth curve, starting in } p \text{ and } U, V \in T_pM.$$

Then we have for an arbitrary $A = \mathcal{R}^{\mathfrak{p}}(U,V) \in \mathfrak{hol}_p(M,h)$ that

$$pr_{\mathfrak{so}(n)} A = pr_E \circ \mathcal{R}^{\mathfrak{p}}(U,V)_{|E} \quad \text{for arbitrary } U, V \in T_pM,$$

i.e. for an $Y \in E$ it is

$$\left(pr_{\mathfrak{so}(n)} A\right) Y = pr_E \left(\mathcal{R}^{\mathfrak{p}}(U,V)Y\right) = \sum_{k=1}^{n} h(\mathcal{R}^{\mathfrak{p}}(U,V)Y, E_k)E_k.$$

For $V_1 = X_1 + Y_1 + Z_1$ and $V_2 = X_2 + Y_2 + Z_2$ with $X_i = \xi_i X$, $Z_i = \zeta_i Z$, $Y_i \in E$, $i = 1,2$ and $Y \in E$ we calculate

$$\begin{aligned}
h(\mathcal{R}^{\mathfrak{p}}(V_1,V_2)Y, E_k) &\overset{\mathfrak{p} \text{ isometry}}{=} h(\mathcal{R}(\mathfrak{p}V_1, \mathfrak{p}V_2)\mathfrak{p}Y, \mathfrak{p}E_k) = \\
&= h(\mathcal{R}(\mathfrak{p}Y_1, \mathfrak{p}Y_2)\mathfrak{p}Y, \mathfrak{p}E_k) + \\
&\quad + h(\mathcal{R}(\mathfrak{p}Z_1, \mathfrak{p}Y_2)\mathfrak{p}Y, \mathfrak{p}E_k) + h(\mathcal{R}(\mathfrak{p}Y_1, \mathfrak{p}Z_2)\mathfrak{p}Y, \mathfrak{p}E_k) \\
&\quad + h(\mathcal{R}(\mathfrak{p}Z_1, \mathfrak{p}X_2)\mathfrak{p}Y, \mathfrak{p}E_k) + h(\mathcal{R}(\mathfrak{p}X_1, \mathfrak{p}Z_2)\mathfrak{p}Y, \mathfrak{p}E_k) \\
&\quad + \underbrace{\underbrace{h(\mathcal{R}(\mathfrak{p}X_1, \mathfrak{p}Y_2)\mathfrak{p}Y, \mathfrak{p}E_k)}_{=-h(\mathcal{R}(\mathfrak{p}Y,\mathfrak{p}E_k)\mathfrak{p}Y_2,\mathfrak{p}X_1)=0} + \underbrace{h(\mathcal{R}(\mathfrak{p}Y_1, \mathfrak{p}X_2)\mathfrak{p}Y, \mathfrak{p}E_k)}_{=h(\mathcal{R}(\mathfrak{p}Y,\mathfrak{p}E_k)\mathfrak{p}Y_1,\mathfrak{p}X_2)=0}}_{\text{because } Y_1, Y_2 \in \Xi^\perp \text{ and the parallel displacements too.}}
\end{aligned}$$

For the terms in the fourth line we have

$$\begin{aligned}
h(\mathcal{R}(\mathfrak{p}X, \mathfrak{p}Z)\mathfrak{p}Y, \mathfrak{p}E_k) &= -h(\mathcal{R}(\mathfrak{p}Y, \mathfrak{p}E_k)\mathfrak{p}Z, \mathfrak{p}X) \\
&= h(\underbrace{\mathcal{R}(\mathfrak{p}E_k, \mathfrak{p}Z)\mathfrak{p}Y}_{\in \Xi^\perp}, \mathfrak{p}X) + h(\underbrace{\mathcal{R}(\mathfrak{p}Z, \mathfrak{p}Y)\mathfrak{p}E_k}_{\in \Xi^\perp}, \mathfrak{p}X) \\
&= 0.
\end{aligned}$$

Hence we obtain

$$\begin{aligned} h(\mathcal{R}^{\mathfrak{p}}(V_1, V_2)Y, E_k) &= \\ &= h(\mathcal{R}(\mathfrak{p}Y_1, \mathfrak{p}Y_2)\mathfrak{p}Y, \mathfrak{p}E_k) + h(\mathcal{R}(\mathfrak{p}Z_1, \mathfrak{p}Y_2)\mathfrak{p}Y, \mathfrak{p}E_k) + h(\mathcal{R}(\mathfrak{p}Y_1, \mathfrak{p}Z_2)\mathfrak{p}Y, \mathfrak{p}E_k) \\ &= h\left(\mathcal{R}\left(\mathfrak{p}Y_1, \mathfrak{p}Y_2\right)\mathfrak{p}Y, \mathfrak{p}E_k\right) + h\left(\mathcal{R}\left(\mathfrak{p}Z, \zeta_1\mathfrak{p}Y_2 - \zeta_2\mathfrak{p}Y_1\right)\mathfrak{p}Y, \mathfrak{p}E_k\right). \end{aligned}$$

Finally it is

$$\begin{aligned} \left(pr_{\mathfrak{so}(n)}A\right)Y &= \sum_{k=1}^{n} h([\mathcal{R}^{\mathfrak{p}}(Y_1, Y_2) + \mathcal{R}^{\mathfrak{p}}(Z, \zeta_1 Y_2 - \zeta_2 Y_1)]Y, E_k)E_k \\ &= pr_E\left(\mathcal{R}^{\mathfrak{p}}(Y_1, Y_2)Y\right) + pr_E\left(\mathcal{R}^{\mathfrak{p}}(Z, \zeta_1 Y_2 - \zeta_2 Y_1)Y\right) \\ &= R(Y_1, Y_2)Y + Q(\zeta_1 Y_2 - \zeta_2 Y_1)Y, \end{aligned}$$

with

$$\begin{aligned} R(.,.) &:= pr_E \circ \mathcal{R}^{\mathfrak{p}}(.,.)_{|E\times E\times E} &\in \Lambda^2 E^* \otimes \mathfrak{g} \\ Q(.) &:= pr_E \circ \mathcal{R}^{\mathfrak{p}}(Z,.)_{|E\times E} &\in E^* \otimes \mathfrak{g}. \end{aligned}$$

Because of the Bianchi identity we have that $R \in \mathcal{K}(\mathfrak{g})$ and $Q \in \mathcal{B}_h(\mathfrak{g})$, as one verifies easily for R

$$\begin{aligned} R(Y_1, Y_2)Y_3 + R(Y_2, Y_3)Y_1 + R(Y_3, Y_1)Y_2 &= \\ pr_E\left(\mathcal{R}^{\mathfrak{p}}(Y_1, Y_2)Y_3 + \mathcal{R}^{\mathfrak{p}}(Y_2, Y_3)Y_1 + \mathcal{R}^{\mathfrak{p}}(Y_3, Y_3)Y_2\right) &= 0, \end{aligned}$$

and for Q

$$\begin{aligned} h(Q(Y_1)Y_2, Y_3) + h(Q(Y_2)Y_3, Y_1) + h(Q(Y_3)Y_1, Y_2) &= \\ pr_E\left(h(\mathcal{R}^{\mathfrak{p}}((Z, Y_1)Y_2, Y_3) + h(\mathcal{R}^{\mathfrak{p}}(Z, Y_2)Y_3, Y_1) + h(\mathcal{R}^{\mathfrak{p}}(Z, Y_3)Y_3, Y_2)\right) &= \\ pr_E\left(h(\mathcal{R}((\mathfrak{p}Z, \mathfrak{p}Y_1)\mathfrak{p}Y_2, \mathfrak{p}Y_3) + h(\mathcal{R}(\mathfrak{p}Z, \mathfrak{p}Y_2)\mathfrak{p}Y_3, \mathfrak{p}Y_1) + h(\mathcal{R}(\mathfrak{p}Z, \mathfrak{p}Y_3)\mathfrak{p}Y_3, \mathfrak{p}Y_2)\right) &= 0. \end{aligned}$$

Here h in $\mathcal{B}_h(\mathfrak{g})$ stands for h restricted to E, i.e. it is positive definite.

This means that the generators of $\mathfrak{g}$ which are $\mathfrak{so}(n)$–projections of the generators of $\mathfrak{hol}_p(M,h)$ are of the form $R(Y_1, Y_2) + Q(Y_3)$ with $Y_i \in E$, $R \in \mathcal{K}(\mathfrak{g})$, $Q \in \mathcal{B}_h(\mathfrak{g})$ for $\mathfrak{g} \subset \mathfrak{so}(E,h)$. With (4.1) from the corollary 4.4 we have $\mathfrak{g} \subset \mathfrak{g}_h$, i.e. $\mathfrak{g}$ is a weak-Berger algebra. □

From point two of proposition 4.5 we get the following

4.7 Corollary. *Let (M,h) be an indecomposable, but non-irreducible Lorentzian manifold and $\mathfrak{g} = pr_{\mathfrak{so}(n)}(\mathfrak{hol}_p(M,h)) \subset E^* \otimes E$ and $\mathfrak{g} = \mathfrak{g}_1 \oplus \ldots \oplus \mathfrak{g}_r$ with $\mathfrak{g}_i \in \mathfrak{so}(E_i, h_i)$ the decomposition in irreducible acting ideals from theorem 2.7. Then these $\mathfrak{g}_i$ are irreducible weak-Berger algebras.*

This corollary ensures that we are at a similar point as in the Riemannian situation, but reaching it by a different way. This is shown schematically in the following diagram:

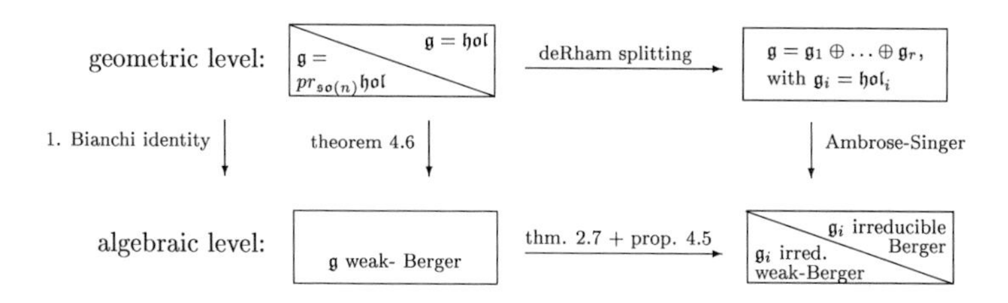

The proof of the theorem gives another

4.8 Corollary. *Let (M,h) be an indecomposable, non-irreducible Lorentzian manifold and $\mathfrak{g} = pr_{\mathfrak{so}(n)}\mathfrak{hol}_p(M,h)$. If there exists coordinates $(x, y_1, \ldots, y_n, z)$ of the above form (i.e. respecting the foliation $\Xi \subset \Xi^{\perp}$), with the property that everywhere holds $\mathcal{R}(\frac{\partial}{\partial z}, \frac{\partial}{\partial y_i}, \frac{\partial}{\partial y_j}, \frac{\partial}{\partial y_k}) = 0$, then $\mathfrak{g}$ is a Berger-algebra.*

The aim of the following sections will be to classify all weak-Berger algebras.

4.1.3 Real and complex weak Berger algebras

Because of the above result we have to classify the real weak-Berger algebras. Since we will use the representation theory of complex semisimple Lie algebras we have to describe the transition of a real weak-Berger algebra to its complexification.
First we note that the spaces $\mathcal{K}(\mathfrak{g})$ and $\mathcal{B}_h(\mathfrak{g})$ for $\mathfrak{g} \subset \mathfrak{so}(E,h)$ can be described by the following exact sequences:

$$\begin{array}{ccccccc} 0 & \to & \mathcal{K}(\mathfrak{g}) & \hookrightarrow & \Lambda^2 E^* \otimes \mathfrak{g} & \xrightarrow{\lambda} & \Lambda^3 E^* \otimes E \\ 0 & \to & \mathcal{B}_h(\mathfrak{g}) & \hookrightarrow & E^* \otimes \mathfrak{g} & \xrightarrow{\lambda_h} & \Lambda^3 E^*, \end{array}$$

where the map λ is the skew-symmetrization and λ_h the dualization by h and the skew-symmetrization.
If we consider a real Lie algebra $\mathfrak{g}$ acting orthogonal on a real vector space E, i.e. $\mathfrak{g} \subset \mathfrak{so}(E,h)$, then h extends by complexification (linear in both components) to a non-degenerate complex-bilinear form $h^{\mathbb{C}}$ which is invariant under $\mathfrak{g}^{\mathbb{C}}$, i.e. $\mathfrak{g}^{\mathbb{C}} \subset \mathfrak{so}(E^{\mathbb{C}}, h^{\mathbb{C}})$. Then the complexification of the above exact sequences gives

$$\mathcal{K}(\mathfrak{g})^{\mathbb{C}} = \mathcal{K}(\mathfrak{g}^{\mathbb{C}}) \tag{4.2}$$
$$(\mathcal{B}_h(\mathfrak{g}))^{\mathbb{C}} = \mathcal{B}_{h^{\mathbb{C}}}(\mathfrak{g}^{\mathbb{C}}). \tag{4.3}$$

This gives:

4.9 Proposition. *$\mathfrak{g} \subset \mathfrak{so}(E,h)$ is a (weak-) Berger algebra if and only if $\mathfrak{g}^{\mathbb{C}} \subset \mathfrak{so}(E^{\mathbb{C}},h^{\mathbb{C}})$ is a (weak-) Berger algebra.*

The *proof* follows directly from the relations (4.2) and (4.3). □

I.e. complexification preserves the weak Berger as well as the Berger property.
Because of proposition 4.5 it suffices to classify the real weak Berger algebras which are irreducible. Now irreducibility is a property which is not preserved under complexification. We have to deal with this problem. At a first step one recalls the following definition, distinguishing two cases for a module of a real Lie algebras.

4.10 Definition. Let $\mathfrak{g}$ be a real Lie algebra. Irreducible real $\mathfrak{g}$-modules E for which $E^{\mathbb{C}}$ is an irreducible $\mathfrak{g}$-module and irreducible complex modules V for which $V_{\mathbb{R}}$ is a reducible $\mathfrak{g}$-module are called of **real type**. Irreducible real $\mathfrak{g}$-modules E for which $E^{\mathbb{C}}$ is a reducible $\mathfrak{g}$-module and irreducible complex modules V for which $V_{\mathbb{R}}$ is a irreducible $\mathfrak{g}$-module are called of **non-real type**.

This notation corresponds to the distinction of complex irreducible $\mathfrak{g}$-modules into real, complex and quaternionic ones. It makes sense because the complexification of a module of real type is of real type — recall that $(E^{\mathbb{C}})_{\mathbb{R}}$ is a reducible $\mathfrak{g}$-module — and the reellification of a module of non-real type is of non-real type. These relations are described in the appendix B.
In the original papers of Cartan [Car14] and Iwahori [Iwa59], see also [Got78], where these distinction is introduced, a representation of real type is called as representation of **first type** and a representation of non-real type is called of **second type**.
If one now complexifies the Lie algebra $\mathfrak{g}$ too, then $E^{\mathbb{C}}$ becomes a $\mathfrak{g}^{\mathbb{C}}$–module. This transition preserves irreducibility.

4.11 Lemma. *Let $\mathfrak{g}^{\mathbb{C}} \subset \mathfrak{gl}(V)$ be the complexification of $\mathfrak{g} \subset \mathfrak{gl}(V)$ with a complex $\mathfrak{g}$-module V. Then it holds:*

1. *$\mathfrak{g}$ is irreducible if and only if $\mathfrak{g}^{\mathbb{C}}$ is irreducible.*
2. *$\mathfrak{g} \subset \mathfrak{so}(V,H)$ if and only if $\mathfrak{g}^{\mathbb{C}} \subset \mathfrak{so}(V,H)$, where H is a symmetric bilinear form.*

Both fact are obvious.
In the following sections we will describe the weak-Berger property for real and non-real modules of a real Lie algebra $\mathfrak{g}$.

4.2 Weak-Berger algebras of real type

In this section we will make efforts to classify weak-Berger algebras of real type, at least the simple ones. The argumentation in this section is analogously to the reasoning in

[MS99].

$\mathfrak{g}_0$ shall be a real Lie algebra and E a real irreducible module of real type. Furthermore we suppose $\mathfrak{g}_0 \in \mathfrak{so}(E,h)$ with h positive definite. Then $E^{\mathbb{C}}$ is an irreducible $\mathfrak{g}_0$-module (also of real type). If $h^{\mathbb{C}}$ denotes the complexification of h, bilinear in both components we have that $\mathfrak{g}_0 \subset \mathfrak{so}(E^{\mathbb{C}}, h^{\mathbb{C}})$.
Now we can extend h also sesqui-linear on $E^{\mathbb{C}}$ and get a hermitian form θ^h on V which is invariant under $\mathfrak{g}_0$. Thus we have $\mathfrak{g}_0 \subset \mathfrak{u}(V, \theta^h)$. θ^h has the same index as h (see appendix B).
Since the bilinear form h we start with is positive definite we can make another simplification. Subalgebras of $\mathfrak{so}(E,h)$ with positive definite h are compact and therefore reductive. I.e. its Levi-decomposition is $\mathfrak{g}_0 = \mathfrak{z}_0 \oplus \mathfrak{d}_0$, with center $\mathfrak{z}_0$ and semisimple derived algebra $\mathfrak{d}_0$. Thus $\mathfrak{g}_0^{\mathbb{C}} = \mathfrak{z} \oplus \mathfrak{d}$ is also reductive. But since it is irreducible by assumption, the Schur lemma implies that the center $\mathfrak{z}$ is $\mathbb{C}\ Id$ or zero. But $\mathbb{C}\ Id$ is not contained in $\mathfrak{so}(V,H)$ and it must be zero. Hence $\mathfrak{g}$ is semisimple. Proposition 4.9 gives the following.

4.12 Proposition. *If $\mathfrak{g}_0 \subset \mathfrak{so}(E,h)$ is a weak-Berger algebra of real type then, $\mathfrak{g}_0^{\mathbb{C}} \subset \mathfrak{so}(E^{\mathbb{C}}, h^{\mathbb{C}})$ is an irreducible weak-Berger algebra. $E^{\mathbb{C}}$ is a $\mathfrak{g}_0$-module of real type and if h is positive definite then $\mathfrak{g}_0^{\mathbb{C}}$ is semisimple.*
If $\mathfrak{g} \subset \mathfrak{so}(V,H)$ is an irreducible complex weak-Berger which is semisimple. Then $\mathfrak{g}$ has a compact real form $\mathfrak{g}_0$ and if V is a $\mathfrak{g}_0$-module of real type, then $V = E^{\mathbb{C}}$, $\mathfrak{g}_0$ is unitary with respect to a hermitian form θ and $\mathfrak{g}_0 \subset \mathfrak{so}(E,h)$ is a weak-Berger algebra of real type. The indices of h and θ are equal.

Proof. The first direction follows obviously from proposition 4.9. That $E^{\mathbb{C}}$ is a module of real type holds because of $(E^{\mathbb{C}})_{\mathbb{R}}$ is reducible (see appendix proposition B.9).
Since $\mathfrak{g}$ is semisimple it has a compact real form $\mathfrak{g}_0$. If V is a $\mathfrak{g}_0$-module of real type, then $\mathfrak{g}_0$ is unitary since it is orthogonal (see proposition B.7) and it is $V = E^{\mathbb{C}}$ (proposition B.9). By proposition B.17 follows that $\mathfrak{g}_0$ is orthogonal w.r.t. h which has the same index as θ. Then the proposition follows by proposition 4.9. □

The main point of this proposition is the implication that if $\mathfrak{g}_0 \subset \mathfrak{so}(E,h)$ is weak-Berger of real type, then $\mathfrak{g}_0^{\mathbb{C}} \subset \mathfrak{so}(E^{\mathbb{C}}, h^{\mathbb{C}})$ is an irreducible acting, complex semisimple weak-Berger algebra. These we have to classify.

4.13 Remark. Before we start we have to make a remark about definition of holonomy up to conjugation. The $SO(n)$–component of an indecomposable, non-irreducible Lorentzian manifold was defined modulo conjugation in $O(n)$. Hence we shall not distinct between subalgebras of $\mathfrak{gl}(n,\mathbb{C})$ which are isomorphic under Ad_{φ} where φ is an element from $O(n,\mathbb{C})$ and Ad the adjoint action in of $Gl(n,\mathbb{C})$ on $\mathfrak{gl}(n,\mathbb{C})$. We say that an orthogonal representation κ_1 of a complex semisimple Lie algebra $\mathfrak{g}$ is **congruent** to an orthogonal representation κ_2 if there is an element $\varphi \in O(n,\mathbb{C})$ such that the following equivalence of $\mathfrak{g}$–representations is valid: $\kappa_1 \sim Ad_\varphi \circ \kappa_2$. Hence we have to classify semisimple, orthogonal, irreducible acting, complex weak-Berger algebras of real type up to this congruence of representations.

If the automorphism Ad_φ is inner, then the representations are equivalent, if it is outer then only congruent.

For semisimple Lie algebras it holds that $Out(\mathfrak{g}) := Aut(\mathfrak{g})/Inn(\mathfrak{g})$ counts the connection components of $Aut(\mathfrak{g})$ and (see for example [OV94]) $Out(\mathfrak{g})$ is isomorphic to the automorphism of the fundamental system, i.e. symmetries of the Dynkin diagram.

For us this becomes relevant in case of $\mathfrak{so}(8,\mathbb{C})$. In the picture one sees that the symmetries of the Dynkin diagram generate the symmetric group $\mathcal{S}_3$, i.e. $Out(\mathfrak{so}(8,\mathbb{C})) = \mathcal{S}_3$ and it contains the so-called "triality automorphism" which interchanges vector and spin representations of $\mathfrak{so}(8,\mathbb{C})$ without fixing one.

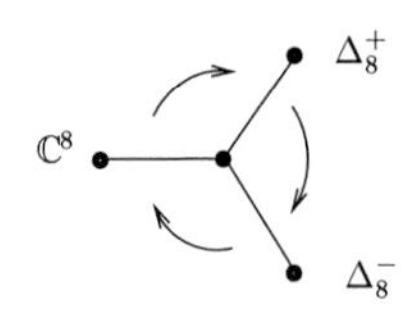

We will use that the automorphism which interchanges the vector representation with one spinor representation and fixes the second spinor representation resp. interchanges the spinor representations and fixes the vector representation comes from Ad_φ with $\varphi \in O(n,\mathbb{C})$. Hence the vector and the spinor representations of $\mathfrak{so}(8,\mathbb{C})$ are congruent to each other.

Finally we should remark that compact real forms equivalent to a given one correspond to inner automorphism of $\mathfrak{g}$. Hence the corresponding representations are equivalent

4.2.1 Irreducible, complex, orthogonal, semisimple Lie algebras

In the following V will be a complex vector space equipped with a non-degenerate symmetric bilinear 2–form H. $\mathfrak{g}$ shall be an irreducible acting, complex, semisimple subalgebra of $\mathfrak{so}(V,H)$.

Thus all the tools of root space decomposition and representation theory will apply. Let $\mathfrak{t}$ be the Cartan subalgebra of $\mathfrak{g}$. We denote by $\Delta \subset \mathfrak{t}^*$ the roots of $\mathfrak{g}$ and we set $\Delta_0 := \Delta \cup \{0\}$. Then $\mathfrak{g}$ decomposes into its root spaces $\mathfrak{g}_\alpha := \{A \in \mathfrak{g} | [T,A] = \alpha(T)\cdot A \text{ for all } T \in \mathfrak{t}\} \neq \{0\}$. It is

$$\mathfrak{g} = \bigoplus_{\alpha\in\Delta_0} \mathfrak{g}_\alpha \quad \text{where } \mathfrak{g}_0 = \mathfrak{t}.$$

By $\Omega \subset \mathfrak{t}^*$ we denote the weights of $\mathfrak{g} \subset \mathfrak{so}(V,H)$. Then V decomposes into the weight spaces $V_\mu := \{v \in V | T(v) = \mu(T)\cdot v \text{ for all } T \in \mathfrak{t}\} \neq \{0\}$, i.e.

$$V = \bigoplus_{\mu\in\Omega} V_\mu.$$

Now the following holds.

4.14 Proposition. *Let $\mathfrak{g} \subset \mathfrak{so}(V, H)$ be a complex, semisimple Lie algebra with weight space decomposition. Then*

$$V(\mu) \perp V(\lambda) \text{ if and only if } \lambda \neq -\mu.$$

In particular if μ is a weight, then $-\mu$ too.

Proof. For any $T \in \mathfrak{t}$, $u \in V_\mu$ and $v \in V_\lambda$ we have

$$0 = H(Tu, v) + H(u, Tv) = (\mu(T) + \lambda(T))\, H(u, v).$$

Now if $\lambda \neq -\mu$ there is a T such that $\mu(T) + \lambda(T) \neq 0$. But this implies $V_\lambda \perp V_\mu$.
On the other hand $V_\mu \perp V_{-\mu}$ would imply $V_\mu \perp V$ which contradicts the non-degeneracy of H.
Its non-degeneracy also implies that $\mu \in \mathfrak{t}^*$ is a weight if and only if $-\mu$ is a weight. □

4.2.2 Irreducible complex weak-Berger algebras

We will now draw consequences from the weak-Berger property. Therefore we consider the space $\mathcal{B}_H(\mathfrak{g})$ defined by the Bianchi identity. If $\mathfrak{g}$ is weak-Berger it has to be non-zero, i.e. by proposition 4.1 it is a non-zero $\mathfrak{g}$–module. If we denote by Π all its weights then it decomposes into weight spaces

$$\mathcal{B}_H(\mathfrak{g}) \;=\; \bigoplus_{\phi \in \Pi} \mathcal{B}_\phi.$$

If Ω are the weights of V then we define the following set

$$\Gamma := \left\{ \mu + \phi \;\middle|\; \begin{array}{l} \mu \in \Omega,\ \phi \in \Pi \text{ and there is an } u \in V_\mu \\ \text{and a } Q \in \mathcal{B}_\phi \text{ such that } Q(u) \neq 0 \end{array} \right\} \subset \mathfrak{t}^*.$$

Then one can prove

4.15 Lemma. $\Gamma \subset \Delta_0$.

Proof. We have to show that every $\mu + \phi \in \Gamma$ is a root of $\mathfrak{g}$. Therefore we consider weight elements $Q_\phi \in \mathcal{B}_\phi$ and $u_\mu \in V_\mu$ with $0 \neq Q_\phi(u_\mu)$. Then for every $T \in \mathfrak{t}$ holds (because of the definition of the $\mathfrak{g}$-module $\mathcal{B}_H(\mathfrak{g})$):

$$\begin{aligned} [T, Q_\phi(u_\mu)] &= (TQ_\phi)(u_\mu) + Q_\phi(T(u_\mu)) \\ &= (\phi(T) + \mu(T))\, Q_\phi(u_\mu) \end{aligned}$$

I.e. $\phi + \mu$ is a root or zero. □

For weak-Berger algebras now the other inclusion is true.

4.16 Proposition. *If $\mathfrak{g} \subset \mathfrak{so}(V,h)$ is irreducible, semisimple Lie algebra. If it is weak-Berger then $\Gamma = \Delta_0$. If $\Gamma = \Delta_0$ and $span\{Q_\mu(u_\mu) \mid \mu \in \Omega\} = \mathfrak{t}$ then it is weak-Berger.*

Proof. The decomposition of $\mathcal{B}_H(\mathfrak{g})$ and V into weight spaces and the fact that $Q_\phi(u_\mu) \in \mathfrak{g}_{\phi+\mu}$ imply the following inclusion

$$\mathfrak{g}_H = span\{Q_\phi(u_\mu) \mid \phi + \mu \in \Gamma\} \subset \bigoplus_{\beta\in\Gamma} \mathfrak{g}_\beta.$$

But if $\mathfrak{g} = \bigoplus_{\alpha\in\Delta_0} \mathfrak{g}_\alpha$ is weak-Berger it holds that $\mathfrak{g} \subset \mathfrak{g}_H$ and thus

$$\bigoplus_{\alpha\in\Delta_0} \mathfrak{g}_\alpha \subset \bigoplus_{\beta\in\Gamma} \mathfrak{g}_\beta \subset \bigoplus_{\alpha\in\Delta_0} \mathfrak{g}_\alpha.$$

But this implies $\Gamma = \Delta_0$.
If now $\Gamma = \Delta_0$ and $span\{Q_\mu(u_\mu) \mid \mu \in \Omega\} = \mathfrak{t}$ we have that

$$\mathfrak{g}_H = span\{Q_\phi(u_\mu) \mid \phi + \mu \in \Gamma\} = \bigoplus_{\beta\in\Gamma} \mathfrak{g}_\beta = \mathfrak{t} \oplus \bigoplus_{\beta\in\Delta} \mathfrak{g}_\beta = \mathfrak{g}.$$

This completes the proof. □

To derive necessary conditions for the weak Berger property we have to fix a notation. Let $\alpha \in \Delta$ be a root. Then we denote by Ω_α the following subset of Ω:

$$\Omega_\alpha := \{\lambda \in \Omega \mid \lambda + \alpha \in \Omega\}.$$

Then of course $\alpha + \Omega_\alpha$ are the weights of $\mathfrak{g}_\alpha V$.

4.17 Proposition. *Let $\mathfrak{g}$ be a semisimple Lie algebra with roots Δ and $\Delta_0 = \Delta \cup \{0\}$. Let $\mathfrak{g} \subset \mathfrak{so}(V,H)$ irreducible, weak-Berger with weights Ω. Then the following properties are satisfied:*

(PI) *There is a $\mu \in \Omega$ and a hyperplane $U \subset \mathfrak{t}^*$ such that*

$$\Omega \subset \{\mu + \beta \mid \beta \in \Delta_0\} \cup U \cup \{-\mu + \beta \mid \beta \in \Delta_0\}. \tag{4.4}$$

(PII) *For every $\alpha \in \Delta$ there is a $\mu_\alpha \in \Omega$ such that*

$$\Omega_\alpha \subset \{\mu_\alpha - \alpha + \beta \mid \beta \in \Delta_0\} \cup \{-\mu_\alpha + \beta \mid \beta \in \Delta_0\}. \tag{4.5}$$

Proof. If $\mathfrak{g}$ is weak-Berger we have $\Gamma = \Delta_0$. We will use this property for $0 \in \Delta_0$ as well as for every $\alpha \in \Delta$.
(PI) $\Gamma = \Delta_0$ implies that there is $\phi \in \Pi$ and $\mu \in \Omega$ such that $0 = \phi + \mu$ with $Q \in \mathcal{B}_\phi$ and $u \in V_\mu$ such that $0 \neq Q(u) \in \mathfrak{t}$, i.e. $\phi = -\mu \in \Pi$. We fix such u, Q and μ. For arbitrary $\lambda \in \Omega$ then occur the following cases:

Case 1: There is a $v_+ \in V_\lambda$ such that $Q(v_+) \neq 0$ or a $v_- \in V_{-\lambda}$ such that $Q(v_-) \neq 0$. This implies $-\mu + \lambda \in \Delta_0$ or $-\mu - \lambda \in \Delta_0$, i.e. $\lambda \in \{\mu + \beta \mid \beta \in \Delta_0\} \cup \{-\mu + \beta \mid \beta \in \Delta_0\}$.

Case 2: For all $v \in V_\lambda \oplus V_{-\lambda}$ holds $Q(v) = 0$. Then the Bianchi identity implies for $v_+ \in V_\lambda$ and $v_- \in V_{-\lambda}$ that $0 = \lambda(Q(u))H(v_+, v_-)$. Now one can choose v_+ and v_- such that $H(v_+, v_-) \neq 0$. This implies $\lambda \in Q(u)^\perp =: U$ and we get (PI).

(PII) Let $\alpha \in \Delta$. $\Gamma = \Delta_0$ implies the existence of $\phi \in \Pi$ and $\mu_\alpha \in \Omega$ such that $\alpha = \phi + \mu_\alpha$ with $Q \in \mathcal{B}_\phi$ and $u \in V_{\mu_\alpha}$ such that $0 \neq Q(u) \in \mathfrak{g}_\alpha$. We fix Q and u for α. This means that $\alpha - \mu_\alpha = \phi \in \Pi$ a weight of $\mathcal{B}_H$.
Let now λ be a weight in Ω_α, i.e. $\lambda + \alpha$ is also a weight. Hence $-\lambda - \alpha$ is a weight. If $v \in V_\lambda$ then $Q(u)v \in V_{\lambda+\alpha}$. Since H is non-degenerate there is a $w \in V_{-\lambda-\alpha}$ such that $H(Q(u)v, w) \neq 0$. Since $Q \in \mathcal{B}_H(\mathfrak{g})$ the Bianchi identity then gives

$$0 = H(Q(u)v, w) + H(Q(v)w, u) + H(Q(w)u, v),$$

i.e. at least one of $Q(v)$ or $Q(w)$ has to be non-zero. Hence we have two cases for $\lambda \in \Omega_\alpha$:

Case 1: $Q(v) \neq 0$. This implies $-\mu_\alpha + \alpha + \lambda \in \Delta_0$, i.e. $\lambda \in \{\mu_\alpha - \alpha + \beta \mid \beta \in \Delta_0\}$.

Case 2: $Q(w) \neq 0$. This implies $-\mu_\alpha + \alpha - \lambda - \alpha = -\mu_\alpha - \lambda \in \Delta_0$, i.e. $\lambda \in \{-\mu_\alpha + \beta \mid \beta \in \Delta_0\}$.

But this is (PII). □

Of course it is desirable to find weights μ and μ_α which are extremal in order to handle criteria (PI) and (PII). To show in which sense this is possible we need a

4.18 Lemma. *Let $\mathfrak{g} \subset \mathfrak{so}(V, H)$ an irreducible, complex semisimple Lie algebra with $\mathcal{B}_H(\mathfrak{g}) \neq 0$. Then for any extremal weight vector $u \in V_\Lambda$ there is a weight element $Q \in \mathcal{B}_H(\mathfrak{g})$ such that $Q(u) \neq 0$.*

Proof. Let $u \in V_\Lambda$ be extremal with $Q(u) = 0$ for every weight element Q. Since $\mathcal{B}_H(\mathfrak{g}) = \bigoplus_{\phi \in \Pi} \mathcal{B}_\phi$ the assumption implies $Q(u) = 0$ for all $Q \in \mathcal{B}_H(\mathfrak{g})$. But this gives for every $A \in \mathfrak{g}$ and every weight element Q that

$$Q(Au) \;=\; [A, Q(u)] - \underbrace{(A \cdot Q)}_{\in \mathcal{B}_H(\mathfrak{g})}(u) \;=\; 0.$$

On the other hand V is irreducible and thats why generated as vector space by elements of the form $A_1 \cdot \ldots \cdot A_k \cdot u$ with $A_i \in \mathfrak{g}$ and $k \in \mathbb{N}$ (see for example [Ser87]). By successive application of $\mathfrak{g}$ to u we get that $Q(v) = 0$ for every weight element Q and every weight vector v. But this gives $Q(v) = 0$ for all $Q \in \mathcal{B}_H(\mathfrak{g})$ and every $v \in V$, hence $\mathcal{B}_H(\mathfrak{g}) = 0$. □

4.19 Proposition. *Let $\mathfrak{g}$ be a semisimple Lie algebra with roots Δ and $\Delta_0 = \Delta \cup \{0\}$. Let $\mathfrak{g} \subset \mathfrak{so}(V,H)$ irreducible, weak-Berger with weights Ω. Then there is an ordering of Δ such that the following holds: If Λ is the highest weight of $\mathfrak{g} \subset \mathfrak{so}(V,H)$ with respect to that ordering, then the following properties are satisfied:*

(QI) *There is a $\delta \in \Delta_+ \cup \{0\}$ and a hyperplane $U \subset \mathfrak{t}^*$ such that*

$$\Omega \subset \{\Lambda - \delta + \beta \mid \beta \in \Delta_0\} \cup U \cup \{-\Lambda + \delta + \beta \mid \beta \in \Delta_0\}. \tag{4.6}$$

If δ cannot be chosen to be zero, then it holds

(QII) *There is an $\alpha \in \Delta$ such that*

$$\Omega_\alpha \subset \{\Lambda - \alpha + \beta \mid \beta \in \Delta_0\} \cup \{-\Lambda + \beta \mid \beta \in \Delta_0\}. \tag{4.7}$$

Proof. First we consider the extremal weights of the representation, i.e. the images of the highest weight under the Weyl group. These do not lie in one hyper plane (because this would imply that all roots lie in one hyperplane). Thus by proposition 4.17 — fixing $\mu \in \Omega$ — there is an extremal weight Λ with $\Lambda + \mu \in \Delta_0$ or $\Lambda - \mu \in \Delta_0$. This one we fix.
Since the Weyl group acts transitively on the extremal weights we can find a fundamental root system, i.e. an ordering on the roots, such that Λ is the highest weight. With respect to this fundamental root system the roots split into positive and negative roots $\Delta = \Delta_+ \cup \Delta_-$. This implies

$$\mu = \varepsilon(\Lambda - \delta) \tag{4.8}$$

with $\alpha \in \Delta_+$ and $\varepsilon = \pm 1$.
For arbitrary $\lambda \in \Omega$ then holds $\lambda \in U = Q(u)^\perp$ or $\lambda + \mu \in \Delta_0$ or $\lambda - \mu \in \Delta_0$. But with (4.8) this implies that we find an $\beta \in \Delta_0$ such that $\lambda = \pm(\Lambda - \delta) + \beta$ with $\beta \in \Delta_0$. This is (QI). Note that we are still free to choose Λ or $-\Lambda$ as highest weight.
Now we suppose that δ cannot be chosen to be zero. Let $v \in V_\Lambda$ be a highest weight vector or $v \in V_{-\Lambda}$. Looking at the proof of proposition 4.17 we have that for all weight elements $Q \in \mathcal{B}_h(\mathfrak{g})$ holds $Q(v) \in \mathfrak{g}_\alpha$ for a $\alpha \in \Delta$. Since $\mathfrak{g}$ is weak-Berger $\mathcal{B}_H(\mathfrak{g})$ is non-zero. Thus we get by lemma 4.18 that there is a weight element Q such that $0 \neq Q(v) \in \mathfrak{g}_\alpha$ and we are done (possibly by making $-\Lambda$ to the highest weight). □

Representations of $\mathfrak{sl}(2,\mathbb{C})$ To illustrate how these criteria shall work we apply them to irreducible representations of $\mathfrak{sl}(2,\mathbb{C})$.

4.20 Proposition. *Let V be an irreducible, complex, orthogonal $\mathfrak{sl}(2,\mathbb{C})$–module of highest weight Λ. If it is weak-Berger then $\Lambda \in \{2,4\}$.*

Proof. Let $\mathfrak{sl}(2,\mathbb{C}) \subset \mathfrak{so}(N,\mathbb{C})$ be an irreducible representation of highest weight Λ. I.e. $\Lambda(H) = l \in \mathbb{N}$ for $\mathfrak{sl}(2,\mathbb{C}) = span(H,X,Y)$ where X has the root α. Since the representation is orthogonal, l must be even (see appendix A) and 0 is a weight. The hypersurface U is the point 0. Now property (4.4) ensures that $l \in \{2,4,6\}$. If $\mu = \Lambda$ we obtain $l \in \{2,4\}$. If $\mu \neq \Lambda$ we can apply (QII): We have that $\Omega_\alpha = \Omega \setminus \{\Lambda\}$ and $\Omega_{-\alpha} = \Omega \setminus \{-\Lambda\}$. Then (QII) implies $l \in \{2,4\}$. □

So we get the first result.

4.21 Corollary. *Let $\mathfrak{su}(2) \subset \mathfrak{so}(E,h)$ be a real irreducible weak-Berger algebra of real type. Then it is a Berger algebra. In particular it is equivalent to the Riemannian holonomy representations of $\mathfrak{so}(3,\mathbb{R})$ on $\mathbb{R}^3$ or of the symmetric space of type AI, i.e. $\mathfrak{su}(3)/\mathfrak{so}(3,\mathbb{R})$ in the compact case or $\mathfrak{sl}(3,\mathbb{R})/\mathfrak{so}(3,\mathbb{R})$ in the non-compact case.*

4.2.3 Berger algebras, weak Berger algebras and spanning triples

In this section we will describe a result of [MS99] and [Sch99], where holonomy groups of torsionfree connections, i.e. Berger algebras, are classified. We will describe our results in their language such that we can use a partial result of [Sch99].
For a Berger algebra holds that for every $\alpha \in \Delta_0$ there is a weight element $R \in \mathcal{K}(\mathfrak{g})$ and weight vectors $u_1 \in V_{\mu_1}$ and $u_2 \in V_{\mu_2}$ such that $0 \neq R(u_1,u_2) \in \mathfrak{g}_\alpha$. The Bianchi identity then gives for an arbitrary $v \in V$

$$R(u_1,u_2)v = R(v,u_2)u_1 + R(u_1,v)u_2.$$

Choosing now u_1, u_2 such that $0 \neq R(u_1,u_2) \in \mathfrak{t}$ one gets for any $\lambda \in \Omega$ and $v \in V_\lambda$ that

$$\lambda(R(u_1,u_2))v = R(v,u_2)u_1 + R(u_1,v)u_2.$$

This implies $\lambda \in (R(u_1,u_2))^\perp \subset \mathfrak{t}^*$ or $V_\lambda \subset \mathfrak{g}V_{\mu_1} \oplus \mathfrak{g}V_{\mu_2}$. This gives property

(RI) There are weights $\mu_1, \mu_2 \in \Omega$ such that

$$\Omega \subset \{\mu_1 + \beta \mid \beta \in \Delta_0\} \cup U \cup \{\mu_2 + \beta \mid \beta \in \Delta_0\}.$$

If one chooses u_1, u_2 such that $0 \neq R(u_1,u_2) = A_\alpha \in \mathfrak{g}_\alpha$ with $\alpha \in \Delta$ then one gets for $\lambda \in \Omega$ that $A_\alpha V_\lambda \subset \mathfrak{g}V_{\mu_1} \oplus \mathfrak{g}V_{\mu_2}$. This means that the weights of $A_\alpha V_\lambda$ are contained in $\{\mu_1 + \beta | \beta \in \Delta_0\} \cup \{\mu_2 + \beta | \beta \in \Delta_0\}$. But this is property

(RII) For every $\alpha \in \Delta$ there are weights $\mu_1, \mu_2 \in \Omega$ such that

$$\Omega_\alpha \subset \{\mu_1 - \alpha + \beta \mid \beta \in \Delta_0\} \cup \{\mu_2 - \alpha + \beta \mid \beta \in \Delta_0\}.$$

Of course our (PI) is a special case of (RI) with $\mu_1 = -\mu_2$. (PII) is not a special case of (RII) since $\mu_\alpha + \alpha$ is not a weight apriori.
To describe this situation further in [Sch99] the following definitions are made. We point out that here Ω_α does not denote the weights of $\mathfrak{g}_\alpha V$ but the weights λ of V such that $\lambda + \alpha$ is a weight.

4.22 Definition. Let $\mathfrak{g} \subset End(V)$ be an irreducible acting complex Lie algebra, Δ_0 be the roots and zero of the semisimple part of $\mathfrak{g}$, Ω the weights of $\mathfrak{g}$ and Ω_α as above.

1. A triple $(\mu_1, \mu_2, \alpha) \in \Omega \times \Omega \times \Delta$ is called **spanning triple** if
$$\Omega_\alpha \subset \{\mu_1 - \alpha + \beta \mid \beta \in \Delta_0\} \cup \{\mu_2 - \alpha + \beta \mid \beta \in \Delta_0\}.$$
2. A spanning triple (μ_1, μ_2, α) is called **extremal** if μ_1 and μ_2 are extremal.
3. A triple (μ_1, μ_2, U) with μ_1, μ_2 extremal weights and U an affine hyperplane in $\mathfrak{t}^*$ is called **planar spanning triple** if every extremal weight different from μ_1 and μ_2 is contained in U and $\Omega \subset \{\mu_1 + \beta \mid \beta \in \Delta_0\} \cup U \cup \{\mu_2 + \beta \mid \beta \in \Delta_0\}$.

From (RI) and (RII) in [Sch99] the following proposition is deduced.

4.23 Proposition. *[Sch99] Let $\mathfrak{g} \subset End(V)$ be an irreducible complex Berger algebra. Then for every root $\alpha \in \Delta$ there is a spanning triple. Furthermore there is an extremal spanning triple or a planar spanning triple.*

If we return to the weak-Berger case we can reformulate proposition 4.19 as follows.

4.24 Proposition. *Let $\mathfrak{g} \subset \mathfrak{so}(V, H)$ be an irreducible complex weak-Berger algebra. Then there is an extremal weight Λ such that one of the following properties is satisfied:*

(SI) *There is a planar spanning triple of the form $(\Lambda, -\Lambda, U)$.*

(SII) *There is an $\alpha \in \Delta$ such that $\Omega_\alpha \subset \{\Lambda - \alpha + \beta \mid \beta \in \Delta_0\} \cup \{-\Lambda + \beta \mid \beta \in \Delta_0\}$.*

There is a fundamental system such that the extremal weight in (SI) and (SII) is the highest weight.

Proof. The proof is analogous the the one of proposition 4.19. If there is an $\alpha \in \Delta$ such that the corresponding μ_α is extremal we are done. If not for every extremal weight vector $u \in V_\Lambda$ and every weight element $Q \in \mathcal{B}_\phi$ holds that $Q(u) \in \mathfrak{t}^*$. Then by lemma 4.18 there is a Q such that $0 \neq Q(u) \in \mathfrak{t}^*$. As before this implies
$$\Omega \subset \{\Lambda + \beta \mid \beta \in \Delta_0\} \cup U \cup \{-\Lambda + \beta \mid \beta \in \Delta_0\}.$$
To ensure that $(\Lambda, -\Lambda, U)$ is a planar spanning triple we have to verify that every extremal weight λ different from Λ and $-\Lambda$ is contained in $U = Q(u)^\perp$. Let λ be

extremal and different from Λ and $-\Lambda$, $v_\pm \in V_{\pm\lambda}$ and $u \in V_\Lambda$. Since $Q(v_\pm) \in \mathfrak{t}$ the Bianchi identity gives

$$\begin{aligned} 0 &= H(Q(u)v_+, v_-) + H(Q(v_+)v_-, u) + H(Q(v_-)u, v_+) \\ &= \lambda(Q(u)) \underbrace{H(v_+, v_-)}_{\neq 0} - \underbrace{\lambda(Q(v_+)) H(v_-, u) + \Lambda((Q(v_-)) H(u, v_+)}_{=0 \text{ since } u \text{ is neither in } V_\lambda \text{ nor in } V_{-\lambda}}. \end{aligned}$$

Hence $\lambda \in U$. □

Obviously we are in a slightly different situation as in the Berger case since $-\Lambda + \alpha$ is not necessarily a weight and in case it is a weight, it is not extremal in general.

4.2.4 Properties of root systems

In this section we will recall the properties of abstract root systems. Let $(E, \langle .,. \rangle)$ be a euclidian vector space. A finite set of vectors Δ is called root system if it satisfies the following properties

1. Δ spans E.

2. For every $\alpha \in \Delta$ the reflection on the hyperplane perpendicular to α defined by

$$s_\alpha(\varphi) := \varphi - \frac{2\langle \varphi, \alpha \rangle}{\|\alpha\|^2}\alpha$$

   maps Δ onto itself.

3. For $\alpha, \beta \in \Delta$ the number $\frac{2\langle \beta, \alpha \rangle}{\|\alpha\|^2}$ is an integer.

A root system is called indecomposable if it does not split into orthogonal subsets. It is called reduced if 2α is not a root if α is a root.
The indecomposable, reduced root systems corresponds to the roots of simple Lie algebras. They are classified in a finite list: A_n, B_n, C_n, D_n, E_6, E_7, E_8, F_4 and G_2. The index designates the dimension of E.
We will cite some basic properties of root systems, which can be found for example in [Kna02].

4.25 Proposition. *(See for example [Kna02], pp. 149) Let Δ be an abstract reduced root system in $(E, \langle .,. \rangle)$.*

1. *If $\alpha \in \Delta$, then the only root which is proportional to α is $-\alpha$.*

2. *If $\alpha, \beta \in \Delta$, then $\frac{2\langle \beta, \alpha \rangle}{\|\alpha\|^2} \in \{0, \pm 1, \pm 2, \pm 3\}$. If Δ is one of the indecomposable root systems ± 3 occurs only for the root system G_2. If both roots are non proportional then ± 2 only occurs for B_n, C_n, F_4 or G_2.*

3. *If α and β are nonproportional in Δ and $\|\beta\| \leq \|\alpha\|$, then* $\frac{2\langle\beta,\alpha\rangle}{\|\alpha\|^2} \in \{0, \pm 1\}$.

4. *Let be $\alpha, \beta \in \Delta$. If $\langle\alpha,\beta\rangle > 0$, then $\alpha - \beta \in \Delta$. If $\langle\alpha,\beta\rangle < 0$, then $\alpha+\beta \in \Delta$. I.e. if neither $\alpha - \beta \in \Delta$ nor $\alpha + \beta \in \Delta$, then $\langle\alpha,\beta\rangle = 0$.*

5. *The subset of Δ defined by $\{\beta + k\alpha \in \Delta \cup \{0\} | k \in \mathbb{Z}\}$ is called α-string through β. It has no gaps, i.e. $\beta + k\alpha \in \Delta$ for $-p \leq k \leq q$ with $p, q \geq 0$ and it holds $p - q = \frac{2\langle\beta,\alpha\rangle}{\|\alpha\|^2}$. The maximal length of such string is given by $\max_{\alpha,\beta\in\Delta} \frac{2\langle\beta,\alpha\rangle}{\|\alpha\|^2} + 1$, i.e. it contains at most four roots.*

As a consequence of that proposition we get the following lemmata. In these we will refer to long and short roots. This notion is evident because in the indecomposable reduced root systems of type B_n, C_n, F_4 and G_2 the roots have two different lengths.

4.26 Lemma. *Let Δ be an indecomposable, reduced root system. Then it holds:*

1. *If $a\alpha + \beta \in \Delta$ for $a \in \mathbb{N}$ and $a > 1$, then $\langle\alpha,\beta\rangle < 0$ and α is a short root.*

2. *If Δ is a root system, where the roots have equal length or if α is a long root, then $\alpha + \beta \in \Delta$ implies $\langle\alpha,\beta\rangle < 0$.*

3. *Let α and β be two short roots. If $\alpha + \beta$ is a long root then $\langle\alpha,\beta\rangle = 0$, if it is a short one then $\langle\alpha,\beta\rangle < 0$. The sum of a short and a long root is a short one*

4. *If β is a long root in $\Delta \neq G_2$, then there are orthogonal roots α and γ such that $\beta = \alpha + \gamma$.*

Proof. The proof follows directly from proposition 4.25. □

4.27 Lemma. *Let α and β be two nonproportional roots and $a \in \mathbb{N}$. If $a(\alpha+\beta) \in \Delta$ then $a = 1$.*

Proof. If $a > 1$ then $\alpha + \beta$ is not a root. This implies $\langle\alpha,\beta\rangle \geq 0$ and yields for $a(\alpha + \beta) = \gamma \in \Delta$:

$$\begin{array}{rcl} 0 < a\left(\|\alpha\|^2 + \langle\alpha,\beta\rangle\right) & = & \langle\alpha,\gamma\rangle \\ 0 < a\left(\langle\alpha,\beta\rangle + \|\beta\|^2\right) & = & \langle\gamma,\beta\rangle. \end{array}$$

On the other hand we have

$$\|\gamma\|^2 = a\left(\langle\alpha,\gamma\rangle + \langle\beta,\gamma\rangle\right).$$

But this gives

$$1 = \frac{a}{2}\underbrace{\Big(\underbrace{\frac{2\langle\gamma,\alpha\rangle}{\|\gamma\|^2}}_{>0 \text{ in } \mathbb{N}} + \underbrace{\frac{2\langle\gamma,\beta\rangle}{\|\gamma\|^2}}_{>0 \text{ in } \mathbb{N}} \Big)}_{\geq 2 \text{ in } \mathbb{N}}.$$

This is a contradiction. Hence $a = 1$. □

The next lemma is a little more general.

4.28 Lemma. *Let α and β be two non-proportional roots in an indecomposable root system and $a, b \in \mathbb{N}$ with $a \leq b$ such that $a\alpha + b\beta \in \Delta$.*

1. *If Δ is not G_2 then $a = 1$. If $\Delta = A_n, D_n, E_6, E_7, E_8$ then $b = 1$ too. If $\Delta = B_n, C_n, F_4$ then $b \leq 2$.*
2. *If $\Delta = G_2$ then $a \leq 2$ and $b \leq 3$.*

Proof. We suppose $a\alpha + b\beta = \gamma \in \Delta$.
First we consider the case $\langle \alpha, \beta \rangle \geq 0$. This gives

$$\begin{array}{lclcl} 0 & < & a\|\alpha\|^2 + b\langle \alpha, \beta \rangle & = & \langle \alpha, \gamma \rangle \\ 0 & < & a(\langle \alpha, \beta \rangle + b\|\beta\|^2 & = & \langle \gamma, \beta \rangle. \end{array}$$

On the other hand we have $\|\gamma\|^2 = a\langle \alpha, \gamma \rangle + b\langle \beta, \gamma \rangle$ and thus

$$\begin{array}{lcl} 1 & = & \frac{a}{2} \overbrace{\frac{2\langle \gamma, \alpha \rangle}{\|\gamma\|^2}}^{>0} + \frac{b}{2} \overbrace{\frac{2\langle \gamma, \beta \rangle}{\|\gamma\|^2}}^{>0} \\ & \geq & \frac{a}{2} \underbrace{\Big(\frac{2\langle \gamma, \alpha \rangle}{\|\gamma\|^2} + \frac{2\langle \gamma, \beta \rangle}{\|\gamma\|^2} \Big)}_{\geq 2 \text{ in } \mathbb{N}}. \end{array}$$

Hence $a = 1$.
Let now be $\langle \alpha, \beta \rangle < 0$. This implies, that $\alpha + \beta =: \delta$ is a root with the property $\delta - \beta = \alpha \in \Delta$
Although the above proposition does not assert that this implies $\langle \delta, \beta \rangle \geq 0$ we can show this. Suppose that $\langle \delta, \beta \rangle < 0$. Hence $\delta + \beta = \alpha + 2\beta$ is a root. If we exclude the root system G_2 point 5 of proposition 4.25 implies $\frac{2\langle \alpha, \beta \rangle}{\|\beta\|^2} = -2$, i.e. $\langle \alpha, \beta \rangle = -\|\beta\|^2$ and finally $\langle \delta, \beta \rangle = 0$, which was excluded.
Thus we have that $\langle \delta, \beta \rangle \geq 0$. Analogously to the first case we get

$$1 = \frac{a}{2} \frac{2\langle \gamma, \delta \rangle}{\|\gamma\|^2} + \frac{(b-a)}{2} \frac{2\langle \gamma, \beta \rangle}{\|\gamma\|^2}.$$

In case that $a \leq b - a$ we get again that $a = 1$. Otherwise we get $b - a = 1$, i.e. $a\delta + \beta = \gamma$. Again by point 5 of proposition 4.25 we get $p - q = \frac{2\langle \gamma, \delta \rangle}{\|\delta\|^2} \geq 0$. But this implies $a \leq 1$.
The possible values for b follow also by proposition 4.25.
For G_2 the possible values of a and b can be calculated analogously. □

4.29 Lemma. *Let η be a long root of an indecomposable root system.*

1. *Let $a, b \in \mathbb{N}$ and $\alpha \in \Delta$ not proportional to η such that $a\eta + b\alpha \in \Delta$. Then $a \leq b$, i.e. $a = 1$ if Δ not equal to G_2 and $a \leq 2$ otherwise.*

2. *Let α, β in Δ not proportional to η and $a \in \mathbb{N}$ such that $a\eta + \alpha + \beta \in \Delta$. Then $a \leq 2$.*

Proof. 1.) First we exclude G_2 and suppose that $b = 1$, i.e. $a\eta + \alpha = \gamma \in \Delta$. Hence $-p \leq a \leq q$ and

$$|p - q| = \frac{2|\langle \eta, \alpha \rangle|}{\|\eta\|^2} < 2\frac{\|\eta\| \cdot \|\eta\|}{\|\eta\|^2} \leq 2,$$

i.e. $|p - q| \leq 1$. But since we have excluded G_2 we have that $a = 1$.
A long root η in G_2 is given by $2e_3 - e_1 - e_2$ with the notations of appendix C of [Kna02]. For this we get the wanted result.
2.) Let $a\eta + \alpha + \beta = \gamma$.
First we consider the case that $\alpha + \beta$ or $\alpha - \gamma$ or $\beta - \gamma$ is a root. If this root is not proportional to η we have by the first point that $a \leq 1$. If it is proportional to η we get that $a \leq 2$ and we are done.
Now we suppose that neither $\alpha + \beta$ nor $\alpha - \gamma$ nor $\beta - \gamma$ is a root. This implies $\langle \alpha, \beta \rangle \geq 0$, $\langle \alpha, \gamma \rangle \leq 0$ and $\langle \beta, \gamma \rangle \leq 0$. We consider the equations

$$\begin{array}{ccccccccc} a\langle \eta, \alpha \rangle & + & \|\alpha\|^2 & + & \underbrace{\langle \alpha, \beta \rangle}_{\geq 0} & = & \langle \gamma, \alpha \rangle & \leq & 0 \\ a\langle \eta, \beta \rangle & + & \|\beta\|^2 & + & \underbrace{\langle \alpha, \beta \rangle}_{\geq 0} & = & \langle \gamma, \beta \rangle & \leq & 0 \\ a\langle \eta, \gamma \rangle & + & \underbrace{\langle \alpha, \gamma \rangle}_{\leq 0} & + & \underbrace{\langle \beta, \gamma \rangle}_{\leq 0} & = & \|\gamma\|^2 & > & 0. \end{array}$$

Hence we have that $\langle \eta, \alpha \rangle < 0$, $\langle \eta, \beta \rangle < 0$ and $\langle \eta, \gamma \rangle > 0$. But since η is a long root, not proportional neither to α nor to β we have that

$$\begin{array}{rcl} \|\eta\|^2 \geq \langle \gamma, \eta \rangle & = & a\|\eta\|^2 + \langle \alpha, \eta \rangle + \langle \beta, \eta \rangle \\ & = & a\|\eta\|^2 - \underbrace{|\langle \alpha, \eta \rangle|}_{<\|\alpha\| \cdot \|\eta\| \leq \|\eta\|^2} - \underbrace{|\langle \beta, \eta \rangle|}_{<\|\beta\| \cdot \|\eta\| \leq \|\eta\|^2} \\ & > & (a-2)\|\eta\|^2. \end{array}$$

This gives $a - 2 < 1$ which is the proposition. □

4.30 Lemma. *Let α be a long root and η be a short one with $\langle \alpha, \eta \rangle > 0$, i.e. $\frac{2\langle \alpha, \eta \rangle}{\|\eta\|^2} \geq 2$. Then there is a short root β with $\beta \not\sim \eta$, $\langle \beta, \alpha \rangle < 0$ and $\langle \beta, \eta \rangle \leq 0$. If the rank of the root system is greater than 2 or if $\frac{2\langle \alpha, \eta \rangle}{\|\eta\|^2} = 3$ (which can only occur for G_2), β can be chosen such that $\langle \beta, \eta \rangle < 0$.*

Proof. $\langle\alpha,\eta\rangle > 0$ implies that $\eta-\alpha$ is a root, in particular a short one. For the inner product we get

$$\langle\alpha,\eta-\alpha\rangle = \langle\alpha,\eta\rangle - \|\alpha\|^2 < \|\alpha\|\|\eta\| - \|\alpha\|^2 < \|\alpha\|^2 - \|\alpha\|^2 = 0$$

and

$$\langle\eta,\eta-\alpha\rangle = \|\eta\|^2 - \langle\alpha,\eta\rangle \leq 0.$$

In case of $\frac{2\langle\alpha,\eta\rangle}{\|\eta\|^2} = 3$ the last $\leq$ is a $<$, and we are done with the second point in case of G_2.
If the rank of the root system is greater than 2 this can be seen with the help of the definitions of the reduced, indecomposable root systems (see appendix C of [Kna02]). □

4.3 Simple complex weak-Berger algebras of real type

In this section we will apply the result of proposition 4.24 to simple complex irreducible acting Lie algebras.
We will do this step by step under the following special conditions:

1. The highest weight of the representation is a root.
2. The representation satisfies (SI), i.e. admits a planar spanning triple $(\Lambda,-\Lambda,U)$.
3. The representation satisfies (SII) and has weight zero.
4. The representation satisfies (SII) and does not have weight zero.

Throughout this section the considered Lie algebra is supposed to be different from $\mathfrak{sl}(2,\mathbb{C})$.

4.3.1 Representations with roots as highest weight

4.31 Proposition. *Let $\mathfrak{g}\subset\mathfrak{so}(N,\mathbb{C})$ be an irreducible representation of real type of a complex simple Lie algebra different from $\mathfrak{sl}(2,\mathbb{C})$ and satisfying (SI) or (SII). If we suppose in addition that there is an extremal weight Λ with $\Lambda = a\eta$ for a root $\eta\in\Delta$ and $a>0$, then the following holds:*

1. *If η is a long root, then $a=1$ and the representation is the adjoint one.*
2. *If η is a short root, then the following holds for a:*
   (a) *If $\Delta = B_n$ or G_2 then $a=1,2$.*
   (b) *If $\Delta = C_n$ or F_4 then $a=1$.*

Proof. Let $\Lambda = a\eta$ with $\eta \in \Delta$, $a \in \mathbb{N}$. W.l.o.g. we may suppose that Λ is the extremal weight in the properties (SI) and (SII). (If not then there is an element of the Weyl group σ mapping Λ to the extremal weight of (SI) and (SII) Λ'. Then $\Lambda' = a\sigma\eta$ and $\sigma\eta \in \Delta$.)
First we show that $a \in \mathbb{N}$. If we chose an fundamental system $(\pi_1, \ldots, \pi_n)$ such that $\Lambda = a\eta$ is the highest weight we get that $\langle\Lambda, \pi_i\rangle = a\langle\eta, \pi_i\rangle \in \mathbb{N}$ for all i. $a \notin \mathbb{N}$ would imply that $\langle\eta, \pi_i\rangle \geq 2$ for all i with $\langle\eta, \pi_i\rangle \neq 0$. This holds only for the root system C_n where $\Lambda = \omega_1 = \frac{1}{2}\eta$. But this representation is symplectic but not orthogonal. (For an explicit formulation of this criterion see [OV94].) So we get $a \in \mathbb{N}$.
Now we consider two cases.

Case 1: η is a long root: In this case the root system of long roots, denoted by Δ_l is the orbit of η under the Weyl group. Hence $a \cdot \Delta_l$ are the extremal weights and $\Delta \subset \Omega$. This implies $0 \in \Omega_\alpha$ for every $\alpha \in \Delta$.

Furthermore for all roots holds that $a \cdot \Delta \subset \Omega$. This is true because we can find a short root such that $\langle\eta, \beta\rangle > 0$. This implies $\eta - \beta \in \Delta_s$. On the other hand it is $\frac{2\langle a\eta, \beta\rangle}{\|\beta\|^2} \geq a$. Hence $a\eta - a\beta = a(\eta - \beta) \in \Omega$. Applying the Weyl group to this weight we get the property for all short roots.

(SI) Let Λ satisfy (SI), i.e. Λ and $-\Lambda$ define a planar spanning triple $(\Lambda, -\Lambda, U)$. This would imply that every long root different from η lies in the hyperplane U. This is only possible for the the root system C_n, because all other root systems have an indecomposable system of long roots. For C_n holds that $\Delta_l = A_1 \times \ldots \times A_1$. But we have still a root β — possibly a short one — such that $\beta \notin U$ and β not proportional to η. This implies $\Omega \ni a\beta = \Lambda + \gamma = a\eta + \gamma$ or $\Omega \ni a\beta = -\Lambda + \gamma = -a\eta + \gamma$ with $\gamma \in \Delta_0$. Then Lemma 4.27 implies $a = 1$.

(SII) Lets suppose that Λ satisfies (SII), i.e. there is an $\alpha \in \Delta$ such that $\Omega_\alpha \subset \{\Lambda - \alpha + \beta | \beta \in \Delta_0\} \cup \{-\Lambda + \beta | \beta \in \Delta_0\}$. $0 \in \Omega_\alpha$ implies $0 = \Lambda - \alpha + \beta = a\eta - \alpha + \beta$ or $0 = -\Lambda + \beta = -a\eta + \beta$ with $\beta \in \Delta_0$. The second is not possible and the first implies by lemma 4.29 that $a = 1$ or $a = 2$ and $\eta = \alpha$. In the second case we find a root $\gamma \not\sim \alpha$ such that $\langle\gamma, \alpha\rangle < 0$, hence $2\gamma \in \Omega_\alpha$. Since $2\gamma - 2\alpha \notin \Delta$ it has to be $2\gamma = \alpha + \beta$, but this is prevented by $\langle\gamma, \alpha\rangle < 0$ and lemma 4.26.

Of course if η is a long root the representation is the adjoint one.

Case 2: η is a short root: Lets denote by Δ_s the root system of short roots. It equals to the orbit of η under the Weyl group. It is a root system of the same rank as Δ

and all roots have the same length. Clearly $\Delta_s \subset \Omega$ and $a \cdot \Delta_s$ are the extremal weights in Ω. For the root system B_n the root system of short roots Δ_s equals to $A_1 \times \ldots \times A_1$, otherwise it is indecomposable.

Furthermore holds the following: If $a \geq 2$ then $\Delta \subset \Omega$. To verify this, we consider a long root $\beta \in \Delta_l$ with the property that $\langle \beta, \eta \rangle > 0$. Such a β always exists. Then we have $\frac{2\langle \eta,\beta\rangle}{\|\eta\|^2} > \frac{2\langle \eta,\beta\rangle}{\|\beta\|^2} \geq 1$. This implies $2\eta - \beta \in \Delta$ (see proposition 4.25). On the other hand $a \geq 2$ ensures that $\Omega \ni s_\beta(2\eta) = 2\left(\eta - \frac{2\langle \eta,\beta\rangle}{\|\beta\|^2}\beta\right)$. This implies that the long root $2\eta - \beta$ is a weight. Now applying the Weyl group to β shows that every long root is a weight.

(SI) We suppose that there is a planar spanning triple $(\Lambda, -\Lambda, U)$. This implies that $a\beta$ lies in the hyperplane U if β is a short root. But this is only possible for B_n because the short roots of all other root systems are indecomposable.

In case of B_n we can at least find a long root α which is not in U. Since the long roots are weights, we have $\alpha = a\eta + \gamma$ or $\alpha = -a\eta + \gamma$ with $\gamma \in \Delta_0$. But this implies for B_n that $a \leq 2$.

(SII) Suppose that there is an $\alpha \in \Delta$ such that $\Omega_\alpha \subset \{\Lambda - \alpha + \beta \mid \beta \in \Delta_0\} \cup \{-\Lambda + \beta \mid \beta \in \Delta_0\}$. $\Delta \subset \Omega$ implies $0 \in \Omega_\alpha$ for all α. $0 = -a\eta + \gamma$ with $\gamma \in \Delta_0$ implies $a = 1$. Hence if we suppose $a \geq 2$ we must have

$$0 = a\eta - \alpha + \gamma \tag{4.9}$$

Thus we have to deal with the following cases:

(a) $\alpha = \eta$ and $a = 2$.

(b) $\alpha \not\sim \eta$ and by 5 of proposition 4.25 $a \leq \frac{2\langle \eta,\alpha\rangle}{\|\eta\|^2} \leq 3$. I.e. if $a \geq 2$, α is a long root.

We exclude the first case for any root system different from B_n. Set $a = 2$ and $\alpha = \eta$. If $\Delta \neq B_n$ the short roots are indecomposable, i.e. there is a short root β such that $\beta \not\sim \eta$ and $\langle \beta, \eta \rangle < 0$. Hence $2\beta \in \Omega_\eta$ and $\beta + \eta \in \Delta$. The existence of a spanning triple implies then $2\beta = \eta + \gamma$ or $2\beta = -2\eta + \gamma$ with $\gamma \in \Delta_0$. The second case is impossible because of lemma 4.27. The first implies $2\beta - \eta \in \Delta$. Again this is not possible by 4.26 and $\langle \beta, \eta \rangle < 0$. Hence the case (a) is excluded.

Now we consider the case (b). First we show that $a = 3$ is not possible. Set $a = 3$. We notice that $\langle \eta, \alpha \rangle > 0$ implies $\frac{2\langle \eta,\alpha\rangle}{\|\alpha\|^2} \geq 1$ and hence $3\eta - 3\alpha \in \Omega_\alpha$. Thus we have the alternative $3\eta - 3\alpha = 3\eta - \alpha + \gamma$ or $3\eta - 3\alpha = -3\eta + \gamma$ with $\gamma \in \Delta_0$. The first implies $2\alpha \in \Delta$ and the second $6\eta - 3\alpha \in \Delta$. Both are not true, hence $a = 3$ is impossible.

We continue with case (b) and have that α is a long root with

$$\frac{2\langle\eta,\alpha\rangle}{\|\eta\|^2} \geq 2 \ , \quad \text{i.e. } 2\eta - \alpha \in \Delta.$$

From now on we suppose, that the root system is different from G_2. Then we have

$$\frac{2\langle\eta,\alpha\rangle}{\|\eta\|^2} = 2. \tag{4.10}$$

In a next step we will show that under these conditions there is no short root β with

$$\beta \in \Delta_s \text{ with } \langle\alpha,\beta\rangle < 0 \ , \ \langle\alpha,\eta\rangle < 0 \text{ and } \beta \not\sim \eta. \tag{4.11}$$

Suppose that there is such a β. Then the first condition implies that $2\beta \in \Omega_\alpha$ and hence $2\beta = 2\eta - \alpha + \gamma$ or $2\beta = -2\eta + \gamma$ with $\gamma \in \Delta_0$. The latter is not possible. The second implies the following using (4.10):

$$-2 \ \geq \ 2 \cdot \frac{2\langle\beta,\eta\rangle}{\|\eta\|^2} \ = \ \frac{2\langle 2\eta-\alpha,\eta\rangle}{\|\eta\|^2} + \frac{2\langle\gamma,\eta\rangle}{\|\eta\|^2} \ = \ 2 + \frac{2\langle\gamma,\eta\rangle}{\|\eta\|^2}.$$

Hence $-4 \geq \frac{2\langle\gamma,\eta\rangle}{\|\eta\|^2}$ which is impossible.
Now by the lemma 4.30 there is such a β. Hence for any remaining root systems different from G_2 and different from B_n we have that $a = 1$.

All in all we have shown, that for a long root holds $a = 1$ and for a short root $a = 2$ implies $\Delta = B_n$ or G_2. □

4.32 Corollary. *Let $\mathfrak{g} \subset \mathfrak{so}(N,\mathbb{C})$ be an irreducible complex simple weak Berger algebra different from $\mathfrak{sl}(2,\mathbb{C})$ and with the additional property that the highest weight is of the form $\Lambda = a\eta$ for a root $\eta \in \Delta$. Then $\mathfrak{g}$ is complexification of a holonomy algebra of a Riemannian manifold or the representation with highest weight $2\omega_1$ of G_2.*

Proof. Clearly if η is a long root the representation is the adjoint one, i.e. the complexification of a holonomy representation of a Lie group with positive definite bi-invariant metric. For a short root η we get the following:

B_n, $a = 1$: This is the representation of highest weight ω_1, i.e. the standard representation of $\mathfrak{so}(2n+1,\mathbb{C})$ on $\mathbb{C}^{2n+1}$. Of course this is the complexification of the generic Riemannian holonomy representation.

B_n, $a = 2$: This is the representation of highest weight $2\omega_1$. A further analysis shows that this is the complexified representation of the Riemannian symmetric space of type AI, i.e. of the symmetric spaces $SU(2n+1)/SO(2n+1,\mathbb{R})$, respectively $SL(2n+1,\mathbb{R})/SO(2n+1,\mathbb{R})$.

C_n, $a = 1$: (for $n \geq 3$) This is the representation of highest weight ω_2. It is the complexified representation of the Riemannian symmetric space of type AII, i.e. of the symmetric spaces $SU(2n)/Sp(n)$, respectively $SL(2n,\mathbb{R})/Sp(n)$.

F_4, $a = 1$: This is the representation of highest weight ω_1. It is the complexified representation of the Riemannian symmetric space of type EIV, i.e. of the symmetric spaces E_6/F_4, respectively $E_{6(-26)}/F_4$.

G_2, $a = 1$: This is the representation of highest weight ω_1. It is the representation of G_2 on $\mathbb{C}^7$, i.e. the complexification of the holonomy of a Riemannian G_2–manifold.

G_2, $a = 2$: This is the representation $2\omega_1$ of G_2. It is a 27-dimensional representation of G_2 isomorphic to $Sym_0^2\mathbb{C}^7$, where $\mathbb{C}^7$ denotes the standard module of G_2 and $Sym_0^2\mathbb{C}^7$ its symmetric, trace free $(2,0)$–tensors. This is the exception, because there is no Riemannian manifold with this complexified holonomy representation.

□

4.3.2 Representations with planar spanning triples

Now we consider representations of a simple Lie algebra under the condition that there is a planar spanning triple. The proof of this proposition is a copy of the proof in [Sch99] adding the additional properties of our planar spanning triple.

4.33 Proposition. *Let $\mathfrak{g} \subset \mathfrak{so}(N,\mathbb{C})$ be an irreducible representation of real type of a complex simple Lie algebra different from $\mathfrak{sl}(2,\mathbb{C})$ and satisfying (SI), i.e. with a planar spanning triple of the form $(\Lambda, -\Lambda, U)$. If there is no root α with $\Lambda = a\alpha$ then $\mathfrak{g}$ is of type D_n with $n \geq 3$ and the representation is congruent to the one with highest weight ω_1 or $2\omega_1$.*

Proof. The condition $\Lambda \neq a\alpha$ implies that there is no root such that $-\Lambda = s_\alpha(\Lambda)$. The existence of a planar spanning triple then gives that for any $\alpha \in \Delta$ such that $\langle \Lambda, \alpha\rangle \neq 0$ the image of the reflection lies in U. If we set $U = T^\perp$ this gives

$$\text{For } \alpha \in \Delta \text{ with } \langle \alpha, \Lambda\rangle \neq 0 \text{ holds} \quad \langle \alpha, T\rangle = \frac{\|\alpha\|^2}{2\langle \Lambda, \alpha\rangle}\langle \Lambda, T\rangle \neq 0. \qquad (4.12)$$

In the following we prove various claims to get the wanted result. We follow completely the lines of reasoning in [Sch99].

Claim 1: *For any non-proportional $\alpha, \beta \in \Delta$ with $\langle \Lambda, \alpha\rangle \neq 0$ and $\langle \Lambda, \beta\rangle \neq 0$ holds that $\langle \alpha, \beta\rangle = 0$ or both have the same length.*

To show this we prove that for two such roots hold that they are orthogonal or that $\langle \Lambda, s_\alpha\beta\rangle = \langle \Lambda, s_\beta\alpha\rangle = 0$.

Suppose that $\langle \Lambda, s_\alpha \beta \rangle \neq 0$. Then (4.12) gives the following

$$\begin{aligned}
\|\beta\|^2 &= \|s_\alpha \beta\|^2 \\
&= \frac{2}{\langle \Lambda, T \rangle} \cdot \langle \Lambda, s_\alpha \beta \rangle \cdot \langle s_\alpha \beta, T \rangle \\
&= \frac{2}{\langle \Lambda, T \rangle} \cdot \left(\langle \Lambda, \beta \rangle - \frac{2\langle \alpha, \beta \rangle}{\|\alpha\|^2} \langle \Lambda, \alpha \rangle \right) \cdot \left(\langle \beta, T \rangle - \frac{2\langle \alpha, \beta \rangle}{\|\alpha\|^2} \langle \alpha, T \rangle \right) \\
&= 2 \cdot \left(\langle \Lambda, \beta \rangle - \frac{2\langle \alpha, \beta \rangle}{\|\alpha\|^2} \langle \Lambda, \alpha \rangle \right) \cdot \left(\frac{\|\beta\|^2}{2\langle \Lambda, \beta \rangle} - \frac{2\langle \alpha, \beta \rangle}{\langle \Lambda, \alpha \rangle} \right) \\
&= 2 \cdot \left(\frac{\|\beta\|^2}{2} - 2\langle \alpha, \beta \rangle \frac{\langle \Lambda, \beta \rangle}{\langle \Lambda, \alpha \rangle} - 2\langle \alpha, \beta \rangle \frac{\langle \Lambda, \alpha \rangle}{\langle \Lambda, \beta \rangle} \frac{\|\beta\|^2}{\|\alpha\|^2} + 4 \frac{2\langle \alpha, \beta \rangle^2}{\|\alpha\|^2} \right).
\end{aligned}$$

Subtracting $\|\beta\|^2$ and multiplying by the denominators gives

$$0 = \langle \alpha, \beta \rangle \left(\|\beta\|^2 \langle \Lambda, \alpha \rangle^2 + \|\alpha\|^2 \langle \Lambda, \beta \rangle^2 - 2\langle \beta, \alpha \rangle \langle \Lambda, \alpha \rangle \langle \Lambda, \beta \rangle \right).$$

But this gives the following pair of equations

$$\begin{aligned}
0 &= \langle \alpha, \beta \rangle \Big(\underbrace{(\|\beta\| \langle \Lambda, \alpha \rangle + \|\alpha\| \langle \Lambda, \beta \rangle)^2}_{>0} - 2 \underbrace{(\|\alpha\| \|\beta\| + \langle \beta, \alpha \rangle)}_{>0} \langle \Lambda, \alpha \rangle \langle \Lambda, \beta \rangle \Big) \\
0 &= \langle \alpha, \beta \rangle \Big(\underbrace{(\|\beta\| \langle \Lambda, \alpha \rangle - \|\alpha\| \langle \Lambda, \beta \rangle)^2}_{>0} + 2 \underbrace{(\|\alpha\| \|\beta\| - \langle \beta, \alpha \rangle)}_{>0} \langle \Lambda, \alpha \rangle \langle \Lambda, \beta \rangle \Big).
\end{aligned}$$

This implies $\langle \alpha, \beta \rangle = 0$ or $\langle \Lambda, \alpha \rangle \langle \Lambda, \beta \rangle = 0$, but this was excluded. This argument is symmetric in α and β hence we get the same result for $s_\beta \alpha$. Thus we have proved that $\langle \Lambda, s_\alpha \beta \rangle = \langle \Lambda, s_\beta \alpha \rangle = 0$ or $\langle \alpha, \beta \rangle = 0$.

Now $\langle \Lambda, s_\alpha \beta \rangle = \langle \Lambda, s_\beta \alpha \rangle = 0$ implies $\langle \Lambda, \alpha \rangle = \frac{2\langle \alpha, \beta \rangle}{\|\alpha\|^2} \cdot \frac{2\langle \alpha, \beta \rangle}{\|\beta\|^2} \cdot \langle \Lambda, \alpha \rangle$. Since $\langle \Lambda, \alpha \rangle$ was supposed to be non zero we have that $\frac{2\langle \alpha, \beta \rangle}{\|\alpha\|^2} \cdot \frac{2\langle \alpha, \beta \rangle}{\|\beta\|^2} = 1$ which implies — since both factors are in $\mathbb{Z}$ — that $\|\alpha\|^2 = \|\beta\|^2$. This holds if $\langle \alpha, \beta \rangle \neq 0$.

Claim 2: *All roots in Δ have the same length.*

Suppose we have short and long roots. Then we can write a long root α as the sum of two short ones, lets say $\alpha = \beta + \gamma$. This implies $\langle \alpha, \beta \rangle \neq 0$ and $\langle \alpha, \gamma \rangle \neq 0$. Since α is long and β and γ are short we have by the first claim that $\langle \Lambda, \alpha \rangle \cdot \langle \Lambda, \beta \rangle = 0$ and $\langle \Lambda, \alpha \rangle \cdot \langle \Lambda, \gamma \rangle = 0$. Now $\langle \Lambda, \alpha \rangle = \langle \Lambda, \beta \rangle + \langle \Lambda, \gamma \rangle$ gives that $\langle \Lambda, \alpha \rangle = 0$ for every long root. But this is impossible. Hence all roots have the same length and in particular holds for non-proportional roots

$$\frac{2\langle \alpha, \beta \rangle}{\|\alpha\|^2} = \pm 1. \tag{4.13}$$

Claim 3: *There is an $a \in \mathbb{N}$ such that for every root α holds $\langle \Lambda, \alpha \rangle \in \{0, \pm a\}$. Furthermore a is less or equal than the length of the roots.*

We consider $\alpha \in \Delta$ with $\langle \Lambda, \alpha \rangle \neq 0$ and set $a := \langle \Lambda, \alpha \rangle$. Then we define the vectorspace $A := span\{\beta \in \Delta \mid \langle \Lambda, \beta \rangle = \pm a\} \subset \mathfrak{t}^*$. We show that $A = \mathfrak{t}^*$ and that every root γ with $\langle \Lambda, \gamma \rangle \notin \{0, \pm a\}$ is orthogonal to A.

To verify $A = \mathfrak{t}^*$ we show that every root is either in A or in $A^\perp$. First consider $\gamma \in \Delta$ with $\langle \Lambda, \gamma \rangle = 0$. If it is not in $A^\perp$ then there are roots $\beta \in A$ and $\delta \notin A$ such that $\gamma = \beta + \delta$. But this implies $0 = \langle \Lambda, \gamma \rangle = \langle \Lambda, \beta \rangle + \langle \Lambda, \delta \rangle = \pm a + \langle \Lambda, \delta \rangle$. Hence $\delta \in A$ and therefore $\gamma \in A$ which is a contradiction. Thus $\gamma \in A^\perp$.

Now we consider a root γ with $\langle \Lambda, \gamma \rangle \notin \{0, \pm a\}$. Then for any β with $\langle \Lambda, \beta \rangle = \pm a$ we have because of (4.13) that $\langle \Lambda, s_\beta \gamma \rangle = \langle \Lambda, \gamma \rangle \pm a \neq 0$. Because of the proof of claim 1 this gives $\langle \beta, \gamma \rangle = 0$. Hence $\gamma \in A^\perp$. Since the root system is indecomposable we have that $A = \mathfrak{t}^*$. Furthermore we have shown that any root with $\langle \Lambda, \gamma \rangle \notin \{0, \pm a\}$ is orthogonal to $A = \mathfrak{t}^*$. Thus the first part of claim 3 is proved.

Now we suppose that $a > c$ where c denotes the length of the roots. We consider an $\alpha \in \Delta$ with $\langle \Lambda, \alpha \rangle = a$. $s_\alpha(\Lambda) = \Lambda - \frac{2a}{c}\alpha$ is an extremal weight in U. Then $a > c$ implies $\Lambda - 2\alpha \in \Omega$ but not in U. Then the existence of the planar spanning triple $(\Lambda, -\Lambda, U)$ implies $\Lambda - 2\alpha = -\Lambda + \beta$ for a $\beta \in \Delta$. Hence

$$\frac{2\langle \Lambda, \gamma \rangle}{c} = 1 + \frac{2\langle \alpha, \gamma \rangle}{c} = 2$$

and therefore $\langle \Lambda, \gamma \rangle = a$ and $a = c$ which is a contradiction.

Now we consider for any $\alpha \in \Delta$ the set $\Delta_\alpha^\perp := \{\beta \in \Delta \mid \langle \alpha, \beta \rangle = 0\} \subset \Delta$. This set is a root system, reduced but not necessarily indecomposable. But we can make the following claim.

Claim 4: *Let $\alpha \in \Delta$ with $\langle \Lambda, \alpha \rangle \neq 0$. Then one of the following cases holds:*

1. *$\Delta_\alpha^\perp$ is orthogonal to Λ or*
2. *there is a unique $\beta \in \Delta_\alpha^\perp$ with $\langle \Lambda, \beta \rangle \neq 0$ such that*
   (a) *$\Lambda = \pm\frac{a}{c}(\alpha + \beta)$ where c is the lengths of the roots, and*
   (b) *$\Delta_\alpha^\perp$ is decomposable with a direct summand $A_1 = \{\pm\beta\}$.*

Suppose that there is a $\beta \in \Delta_\alpha^\perp$ with $\langle \Lambda, \beta \rangle \neq 0$. W.l.o.g. we can suppose that $\langle \Lambda, \beta \rangle = \langle \Lambda, \alpha \rangle = \pm a$. $\langle \alpha, \beta \rangle$ implies then

$$s_\alpha s_\beta(\Lambda) = \Lambda \mp \frac{2a}{c}(\alpha + \beta).$$

Now we show with the help of (4.12) that $s_\alpha s_\beta(\Lambda)$ is not in U:

$$\begin{aligned} \langle s_\alpha s_\beta(\Lambda), T \rangle &= \langle \Lambda, T \rangle - \frac{2\langle \Lambda, \alpha \rangle}{\|\alpha\|^2}\langle \alpha, T \rangle - \frac{2\langle \Lambda, \beta \rangle}{\|\beta\|^2}\langle \beta, T \rangle \\ &= -\langle \Lambda, T \rangle \neq 0. \end{aligned}$$

But this implies $-\Lambda = s_\alpha s_\beta(\Lambda) = \Lambda \pm \frac{2a}{c}(\alpha + \beta)$. By this equation α determines β uniquely.

We still have to show that such β is orthogonal to all other roots in Δ_α. For $\gamma \not\perp \beta$ in Δ_α we have

$$\langle \Lambda, s_\beta \gamma \rangle = \underbrace{\langle \Lambda, \gamma \rangle}_{=0} - \frac{2\langle \beta, \gamma \rangle}{\|\beta\|^2} \langle \Lambda, \beta \rangle.$$

The uniqueness of β implies that β is orthogonal to Δ_α.

Claim 5: *The root system of $\mathfrak{g}$ is of type A_n or D_n.*

The only root system with roots of equal length where the root system $\Delta_\alpha^\perp$ is decomposable for a root α is D_n. Hence for every root system different from D_n we have that $\Delta_\alpha^\perp \perp \Lambda$ by claim 4. Any root system different from A_n satisfies that $span(\Delta_\alpha^\perp) = \alpha^\perp$. Both together imply that for any root system different from D_n and A_n we have that $\alpha = \Lambda$ but this was excluded.

To find the representations of A_n and D_n which obey the above claims we introduce a fundamental system $\Pi = (\pi_1, \ldots, \pi_n)$ which makes Λ to the highest weight of the representation. Then we have that $\Lambda = \sum_{k=1}^n m_k \omega_k$ with $m_k \in \mathbb{N} \cup \{0\}$ and ω_k the fundamental representations. $\langle \omega_i, \pi_j \rangle = \delta_{ij}$ implies $m_i = \langle \Lambda, \pi_i \rangle \in \{0, a\}$. Then we get

Claim 6. *The root system is of type D_n and the representation is the a-th power of a fundamental representation, i.e. $\Lambda = a\omega_i$.*

Applying Λ to the root $\sum_{k=1}^n \pi_k$ gives $\sum_{k=1}^n m_k = a$. Applying Λ to any of the π_i gives that $\sum_{k=1}^n m_k = m_i$ for one i.

Now we consider the root system A_n. $n = 1$ was excluded from the beginning. Recalling $A_3 \simeq D_3$ we can also exclude A_3. Now we impose the condition that the representation is orthogonal. This forces n to be odd and $\Lambda = a\omega_{\frac{n+1}{2}}$ where a has to be 2 when $\frac{n+1}{2}$ is odd. Thus we can suppose that $n > 3$. Using the notation of the appendix we consider now the root $\sum_{k=1}^n \pi_k = e_1 - e_{n+1}$ for which holds that $\langle \Lambda, \eta \rangle = a$. Hence by claim 4 we have that $\Delta_\eta^\perp$ is orthogonal to Λ. On the other hand $\Delta_\eta^\perp = \{\pm(e_i - e_j) \mid 2 \leq i < j \leq n\}$ with $n > 3$ is not orthogonal to $a\omega_{\frac{n+1}{2}} = a\left(e_1 + \ldots + e_{\frac{n+1}{2}}\right)$. This yields a contradiction.

Finally we show that only the representations of D_n given in the proposition satisfy the derived properties. The fundamental representations of D_n are given by $\omega_i = e_1 + \ldots + e_i$ for $i = 1 \ldots n-2$ and $\omega_i = \frac{1}{2}(e_1 + \ldots + e_{n-1} \pm e_n)$ for $i = n-1, n$. Then $\langle a\omega_i, \pi_i \rangle = a$. On the other hand for the largest root $\eta = e_1 + e_2$ holds

$$\langle a\omega_i, \eta \rangle = \begin{cases} a & : \ i = 1, n-1, n \\ 2a & : \ 2 \leq i \leq n-2. \end{cases}$$

Hence the representation of $a\omega_i$ with $2 \leq i \leq n-2$ does not satisfy claim 3. Now we consider for $n > 4$ the representations $\Lambda = \frac{1}{2}(e_1 + \ldots + e_{n-1} \pm e_n)$. For the root $\alpha = e_{n-1} \pm e_n$ holds that $\langle \Lambda, \alpha \rangle = a \neq 0$. The roots $\beta := e_1 - e_2$ and $\gamma := e_1 + e_3$ both satisfy $\langle \Lambda, \beta \rangle = \langle \Lambda, \gamma \rangle = a$ and $\langle \alpha, \beta \rangle = \langle \alpha, \gamma \rangle = 0$. But this is a violation of the uniqueness property in claim 4. Hence $n = 4$.
For D_4 it holds that, ω_3 and w_4 are congruent to ω_1, i.e. there is an involutive automorphism of the Dynkin diagram which interchanges ω_1 with ω_3 respectively ω_1 with ω_4. For $D_3 \simeq A_3$ only the representations ω_2 and $2\omega_2$ are orthogonal. □

Again we get a

4.34 Corollary. *Every representation of a Lie algebra which satisfies the conditions of proposition 4.33 is the complexification of a Riemannian holonomy representation.*

Proof. The representation with highest weight ω_1 of D_n is the standard representation of $\mathfrak{so}(2n, \mathbb{C})$ in $\mathbb{C}^{2n}$. Hence it is the holonomy representation of a generic Riemannian manifold.
The representation with highest weight $2\omega_1$ is the complexified holonomy representation of a symmetric space of type AI for even dimensions, i.e. of $SU(2n)/SO(2n, \mathbb{R})$ respectively $Sl(2n, \mathbb{R})/SO(2n, \mathbb{R})$. □

4.3.3 Representations with the property (SII) and weight zero

Now we will study the property (SII) for representation with weight zero. For this we need a lemma.

4.35 Lemma. *Let $\mathfrak{g} \subset \mathfrak{so}(N, H)$ the irreducible representation of a simple Lie algebra with weights Ω. If $0 \in \Omega$ then*

1. *$\Delta \subset \Omega$ or*
2. *the extremal weights are short roots or*
3. *$\Delta = C_n$ and the representation is a fundamental one with highest weight ω_{2k} for $k \geq 2$.*

Proof. $0 \in \Omega$ implies that there is a $\lambda \in \Omega$ and an $\eta \in \Delta$ such that $0 = \lambda - \eta$, i.e $\lambda = \eta$. Now we consider two cases.

Case 1: η is a long root.

Of course we have that the root system of long roots is contained in Ω. We have to show that the short roots are in Ω. This is the case if one short root is in Ω. For this we write $\eta = \alpha + \beta$ where α and β are short roots. If $\Delta \neq G_2$ we have that

$\langle\alpha,\beta\rangle = 0$. In this case we have that $\frac{2\langle\eta,\alpha\rangle}{\|\alpha\|^2} = 2$, i.e.$\eta - \alpha = \beta \in \Omega$. For $\Delta = G_2$ we have that $\frac{2\langle\alpha,\beta\rangle}{\|\alpha\|^2} = \frac{2\langle\alpha,\beta\rangle}{\|\beta\|^2} = 1$ and therefore $\frac{2\langle\eta,\alpha\rangle}{\|\alpha\|^2} = 3$, i.e.$\eta - \alpha = \beta \in \Omega$ too. Hence also the short roots are weights and we have $\Delta \subset \Omega$.

Case 2: η is a short root.

Again the short roots are weights. We have to show that one long root is a weight if η is not extremal or that we are in the case of the C_n with the above representations. If η is not extremal then exists an $\alpha \in \Delta$ such that $\eta + \alpha \in \Omega$ and $\eta - \alpha \in \Omega$. This α we fix and consider the following cases.

Case A: $\alpha = \eta$, i.e. $2\eta \in \Omega$. If $\Delta \neq G_2$ we find a long root β such that $\frac{2\langle\eta,\beta\rangle}{\|\eta\|^2} = -2$. This implies that $\beta + 2\eta$ is a long root but also a weight. In case of G_2 we find a short root β with $\langle\eta,\beta\rangle < 0$ and such that $2\eta + \beta \in \Delta$ a long root. This long root is also in Ω since $\langle\eta,\beta\rangle < 0$.

Case B: $\alpha \not\sim \eta$ and $\langle\alpha,\eta\rangle \neq 0$. First we consider the case where α is a short root. W.l.o.g. let be $\langle\alpha,\eta\rangle < 0$ Then $\alpha+\eta$ is a root and a weight. If Δ is different from C_n it is a long root and we are ready. For C_n we have to analyze the situation in detail (see the appendix of [Kna02]): Let $\eta = e_i + e_j$ and $\alpha = e_k - e_j$ with $i \neq k$ be the two short roots. Since $\Omega \ni \eta - \alpha = 2e_j + e_i - e_k$ we have that $\frac{2\langle\eta-\alpha,e_i-e_k\rangle}{\|e_i-e_k\|^2} = 2$. Hence $\eta - \alpha - (e_i - e_k) = 2e_j \in \Omega$. But $2e_j$ is a long root of C_n and we are ready.

If α is a long root we proceed as follows. For G_2 one of $\eta \pm \alpha$ is a short root, lets say $\eta - \alpha$. Then we have that $\langle\eta+\alpha,\eta-\alpha\rangle < 0$ hence 2η is a weight and we may argue as in the first case A. If Δ is different from G_2 we write $\alpha = \alpha_1 + \alpha_2$ with two orthogonal short roots α_1 and α_2. For one of these is $\langle\eta,\alpha\rangle \neq 0$ and hence $\eta \pm \alpha_i$ a long root, but also a weight.

Case C: $\langle\alpha,\eta\rangle = 0$ and $\Delta \neq C_n$. For G_2 this case implies that α is a long root and that $\eta + \alpha$ is two times a short root. Then for G_2, we can proceed as above to get the result.

If Δ is different from G_2 we consider the root system $\Delta_\eta^\perp$ of roots orthogonal to η which contains α. In case of C_n this root system is equal to $A_1 \times C_{n-2}$ and in the remaining cases — B_n and F_4 — equal to B_{n-1} resp. B_3. Now we show that there is a short root α_1 in $\Delta_\eta^\perp$ such that $\eta + \alpha_1 \in \Omega$. If α is short this is trivial and if α is long we write $\alpha = \alpha_1 + \alpha_2$ with two orthogonal short roots from $\Delta_\eta^\perp$. Then $\langle\eta+\alpha,\alpha_2\rangle > 0$ and thus $\eta + \alpha_1 \in \Omega$.

On the other hand there is a short root $\gamma \in \Delta_\eta^\perp$ with $\eta + \gamma$ is a long root. Applying now the Weyl group of $\Delta_\eta^\perp$ on $\eta + \gamma$ we get that $\eta + \alpha_1$ is a long root. In case of C_n this argument does not apply since γ spans the A_1 factor of $\Delta_\eta^\perp$.

Hence we have verified $\Delta \subset \Omega$ in the cases A, B and C. It remains to show that in the situation where $\langle \alpha, \eta \rangle = 0$, $\Delta = C_n$ and neither case A nor case B applies, it holds that $\Delta \subset \Omega$ or the representation of C_n is the one with highest weight ω_{2k} with $k \geq 2$.

We suppose that $\Delta \not\subset \Omega$. Hence no long root can be a weight.

First of all we show that under these conditions α has to be a short root. This is true because $\frac{2\langle \eta \pm \alpha, \eta \rangle}{\|\eta\|^2} = 2$ implies $\Omega \ni \eta \pm \alpha - \eta = \pm\alpha$. Hence α has to be short.

Secondly we note that neither $\eta + \alpha$ nor $\eta - \alpha$ can be a root because it would be a long root and a weight. This implies $n \geq 4$.

In a third step we show that there is no long root β such that $\eta + \alpha + \beta \in \Omega$ and $\eta + \alpha - \beta \in \Omega$. We consider the number

$$\frac{2\langle \eta + \alpha \pm \beta, \alpha \rangle}{\|\alpha\|^2} = 2 \pm \frac{2\langle \alpha, \beta \rangle}{\|\alpha\|^2}. \tag{4.14}$$

If $\langle \alpha, \beta \rangle = 0$ we have that $\eta + \alpha \pm \beta - \alpha = \eta \pm \beta \in \Omega$. But this was excluded (First step or case B). Hence we suppose that $\frac{2\langle \alpha, \beta \rangle}{\|\alpha\|^2} = 2$. We still have that $\eta + \beta \in \Omega$. We consider the number $\frac{2\langle \eta + \beta, \eta \rangle}{\|\eta\|^2} = 2 \pm \frac{2\langle \eta, \beta \rangle}{\|\eta\|^2} \geq 0$. If this is not zero we have that $\Omega \ni \eta + \beta - \eta = \beta$ which was excluded. Hence $\frac{2\langle \eta, \beta \rangle}{\|\eta\|^2} = -2$. But this together with $\frac{2\langle \alpha, \beta \rangle}{\|\alpha\|^2} = 2$ is a contradiction since the long roots of C_n are of the form $\pm 2e_i$ and the short ones of the form $\pm(e_i \pm e_j)$. Hence if there is a root such that $\eta + \alpha + \beta \in \Omega$ and $\eta + \alpha - \beta \in \Omega$, it has to be a short one.

If there is no such β then $\eta + \alpha$ is extremal. Looking at the fundamental weights of C_n one sees that this implies that the highest weight of the representation is ω_4.

Finally we suppose that there is such a short root β. Since β is short equation (4.14) implies $\eta \pm \beta \in \Omega$. Since we have excluded case A and B it has to be $\langle \eta, \beta \rangle = 0$ and neither $\eta + \beta$ nor $\eta - \beta$ is a root. On the other hand the same holds for α and β since any other would imply that $\alpha \pm \beta$ is a long root which was excluded or a short root γ orthogonal to η and with $\eta \pm \gamma \in \Omega$. This way we go on attaining that any extremal wight is the sum of orthogonal short roots whose pairwise sum is no long root. But this is nothing else than the fact that the highest weight of the representation is ω_{2k} for $k \geq 2$.

All in all we have shown the proposition. □

4.36 Proposition. *Let $\mathfrak{g} \subset \mathfrak{so}(N, \mathbb{C})$ be an irreducible representation of real type of a complex simple Lie algebra different from $\mathfrak{sl}(2, \mathbb{C})$ and satisfying (SII). If $0 \in \Omega$ then there is a root α such that for the extremal weight from property (SII) holds $\Lambda = a\alpha$ or the representation is congruent to one of the following:*

1. *$\Delta = C_4$ with highest weight ω_4.*
2. *$\Delta = D_n$ with highest weight $2\omega_1$.*

Proof. Let Λ and α be the extremal weight and the root from property (SII). We suppose that Λ is not the multiple of a root.
First of all we consider the case where $0 \in \Omega_\alpha$. By the previous lemma this is true in the following cases:

(a) $\Delta \neq C_n$, because in this case $\Delta \subset \Omega$.

(b) $\Delta = C_n$ but the highest weight of the representation is not equal to ω_{2k} with $k \geq 2$, because this again implies $\Delta \subset \Omega$.

(c) $\Delta = C_n$ and α is a short root, because for representations with $0 \in \Omega$ holds that the short roots are weights.

For $0 \in \Omega_\alpha$ property (SII) gives that $0 = \Lambda - \alpha - \beta$ or $0 = -\Lambda + \beta$. The second case was excluded thus we have to consider the first case. Suppose that $\Lambda = \alpha + \beta$ where $\alpha + \beta \not\sim \gamma \in \Delta$. In particular $\alpha + \beta$ is not a root which implies that $\langle \alpha, \beta \rangle \geq 0$. We consider three cases.

Case 1: $\Delta = G_2$. In this case $\alpha + \beta \not\sim \gamma \in \Delta$ implies $\langle \alpha, \beta \rangle > 0$ and α and β must have different length. Thus we can chose a long root γ not proportional neither to α nor to β and such that $\langle \alpha, \gamma \rangle < 0$ and $\langle \beta, \gamma \rangle < 0$ which implies $\gamma \in \Omega_\alpha$ as well as $\gamma \in \Omega_\beta$. (SII) implies then $\gamma - \beta \in \Delta$ or $\gamma - \alpha \in \Delta$ or $\gamma + \alpha + \beta \in \Delta$. The first two cases are not possible because of lemma 4.26. For the third case we suppose that α is the long root and consider $\frac{2\langle \gamma+\beta, \alpha \rangle}{\|\alpha\|^2} = 0$ because α is long and both terms have opposite sign. Hence $\gamma + \alpha + \beta$ cannot be a root.

Case 2: $\Delta \neq G_2$ *and* $\langle \alpha, \beta \rangle > 0$. This implies $\alpha - \beta \in \Delta$. We consider the number $k := \frac{2\langle \alpha, \alpha+\beta \rangle}{\|\alpha\|^2} = 2 + \frac{2\langle \alpha, \beta \rangle}{\|\alpha\|^2} \geq 3$. Since G_2 was excluded we have that $k \in \{3, 4\}$. Hence $\alpha + \beta - k\alpha = \beta - (k-1)\alpha \in \Omega_\alpha$. Then property (SII) implies $\beta - (k-1)\alpha = -\alpha - \beta + \gamma$ with $\gamma \in \Delta_0$, i.e. $2\beta - (k-2)\alpha \in \Delta$. At first this implies $k = 3$ and thus $\frac{2\langle \alpha, \beta \rangle}{\|\alpha\|^2} = 1$. Secondly we must have $\frac{2\langle \alpha, \beta \rangle}{\|\beta\|^2} = 2$, therefore $\|\alpha\|^2 = 2\|\beta\|^2$, i.e. α as well as $2\beta - \alpha$ are long roots and β and $\beta - \alpha$ are short ones.

This implies $\frac{2\langle \beta-\alpha, \alpha+\beta \rangle}{\|\beta-\alpha\|^2} = \frac{2(\|\beta\|^2 - \|\alpha\|^2)}{\|\beta\|^2} = -2$. Hence $\alpha + \beta + 2(\beta - \alpha) = 3\beta - \alpha \in \Omega$ and since $\frac{2\langle \alpha, \alpha-3\beta \rangle}{\|\alpha\|^2} = 2 - 3 = -1$ holds $\alpha - 3\beta \in \Omega_\alpha$. (SII) then gives $\alpha - 3\beta = \beta - \gamma$ or $\alpha - 3\beta = -\beta - \alpha + \gamma$ with $\gamma \in \Delta_0$. But none of these equations can be true.

Case 3: $\langle\alpha,\beta\rangle = 0$ *and* $\Delta \neq G_2$. Since $\alpha+\beta \not\sim \gamma \in \Delta$ the rank of Δ has to be greater than 3 or it is $\Delta = D_n$ and $\Lambda = 2e_i$, i.e. $\Lambda = 2w_1$. In the second case we are ready and we exclude this representation in the following. We can suppose $rk\Delta \geq 4$. In this situation we prove the following lemma.

4.37 Lemma. *Let* $rk\Delta \geq 4$ *and let* $\Lambda = \alpha+\beta$ *be an extremal weight of a representation satisfying property (SII) for* $(\Lambda, -\Lambda+\alpha, \alpha)$ *with* $\beta \in \Delta$ *satisfying* $\langle\alpha,\beta\rangle = 0$ *and* $\alpha+\beta \not\sim \gamma \in \Delta$. *Then* Δ *is a root system with roots of the same length or* $\Delta = C_n$ *and* α *and* β *are two short roots.*

Proof. Suppose that Δ has roots of different length.

First we assume that β is a long root. We consider the root system $\Delta_\alpha^\perp$, which contains β. We note that β lies not in the A_1 factor of $\Delta_\alpha^\perp$ because otherwise $\alpha+\beta$ would be the multiple of a root. Since β is long we find a short root $\gamma \in \Delta_\alpha^\perp$ such that $\frac{2\langle\beta,\gamma\rangle}{\|\gamma\|^2} = -2$. Hence $\alpha+\beta+2\gamma \in \Omega$ and — since $\frac{2\langle\alpha,\alpha+\beta+2\gamma\rangle}{\|\alpha\|^2} = 2$ — it is $-\alpha-\beta-2\gamma \in \Omega_\alpha$. But this contradicts property (SII).

Now we suppose that α is a long root. Here we consider the root system $\Delta_\beta^\perp$ containing α. Again α lies not in the A_1 factor of $\Delta_\beta^\perp$ because otherwise $\alpha+\beta$ would be the multiple of a root. Since α is long we find a short root $\gamma \in \Delta_\beta^\perp$ such that $\frac{2\langle\alpha,\gamma\rangle}{\|\gamma\|^2} = -2$. Hence $\alpha+\beta+2\gamma \in \Omega$. Now we have that $\frac{2\langle\alpha,\gamma\rangle}{\|\alpha\|^2} = -1$ and therefore $\frac{2\langle\alpha,\alpha+\beta+2\gamma\rangle}{\|\alpha\|^2} = 2-1 = 1$. Thus $-\alpha-\beta-2\gamma \in \Omega_\alpha$. Again this contradicts (SII).

If α and β are short and orthogonal and the root system is not C_n, i.e. it is B_n or F_4, then the sum of two orthogonal short roots is the multiple of a root. □

Now we prove a second

4.38 Lemma. *The assumptions of the previous lemma imply that there is no* $\gamma \in \Delta$ *such that*

$$\langle\alpha,\gamma\rangle = 0 \text{ and } \frac{2\langle\beta,\gamma\rangle}{\|\gamma\|^2} = 1. \tag{4.15}$$

Proof. Lets suppose that there is a $\gamma \in \Delta$ such that $\langle\alpha,\gamma\rangle = 0$ and $\frac{2\langle\beta,\gamma\rangle}{\|\gamma\|^2} = 1$. In case of C_n γ is a short root. We note that both together imply that neither $\alpha+\gamma$ nor $\alpha-\gamma$ is a root. But $\gamma-\beta$ is a root, in case of C_n a short one. Furthermore $\Lambda-\gamma \in \Omega$ Hence

$$\frac{2\langle\Lambda-\gamma,\gamma-\beta\rangle}{\|\gamma-\beta\|^2} = \frac{2\langle\alpha+\beta-\gamma,\gamma-\beta\rangle}{\|\gamma-\beta\|^2} = -2.$$

Hence $\Lambda-\gamma+2(\gamma-\beta)=\alpha-\beta+\gamma\in\Omega$. Now $\frac{2\langle\alpha-\beta+\gamma,\alpha\rangle}{\|\alpha\|^2}=2$, i.e. $-\alpha+\beta-\gamma\in\Omega_\alpha$. (SII) implies now that $-\alpha+\beta-\gamma=\beta+\delta$ or $-\alpha+\beta-\gamma=-\alpha-\beta+\delta$ for $\delta\in\Delta_0$. But both options are not possible since $\alpha+\gamma$ is not a root and because γ is short. □

We conclude that lemma 4.37 left us with representations of A_n, D_n, E_6, E_7, E_8 or C_n where Λ is the sum of two orthogonal (short) roots but not a root.

Now one easily verifies that lemma 4.38 implies $n\leq 4$ and $\Delta\neq A_4$. Hence the remaining representations are $2\omega_1$, $2\omega_3$ and $2\omega_4$ of D_4, which are congruent to each other, and w_4 of C_4.

To finish the proof we have to consider the representation of highest weight ω_{2k} (with $k\geq 2$) of C_n supposing α is a long root. $0\in\Omega$ implies that the short roots are weights. Let β be a short root with $\langle\alpha,\beta\rangle<0$, i.e. $\beta\in\Omega_\alpha$. (SII) then gives $\beta=\omega_{2k}-\alpha+\delta$ or $\beta=\omega_{2k}-\delta$ for a $\delta\in\Delta_0$. Analyzing roots and fundamental weights of C_n we get that (SII) implies $k=2$ and $\alpha=2e_i$ for $1\leq i\leq 4$. But for $n>4$ lemma 4.38 applies analogously. The remaining representation is ω_4 of C_4. □

4.39 Corollary. *Let $\mathfrak{g}\subset\mathfrak{so}(N,\mathbb{C})$ be an orthogonal algebra of real type different from $\mathfrak{sl}(2,\mathbb{C})$ and satisfying (SII). If $0\in\Omega$, in particular if $\Delta=G_2,F_4$ or E_8 then it is the complexification of a Riemannian holonomy representation with the exception of G_2 in corollary 4.32.*

Proof. If Λ is the multiple of a root then we are in the situation of corollary 4.32. For D_n the remaining representations are those which appear in corollary 4.34. The representation of highest weight ω_4 of C_4 is the complexification of the holonomy representation of the Riemannian symmetric space of type EI, i.e. of $E_6/Sp(4)$ resp. $E_{6(6)}/Sp(4)$. Furthermore analyzing the roots and fundamental representations of the exceptional algebras we notice that every representation of G_2, F_4 and E_8 contains zero as weight. □

4.3.4 Representations with the property (SII) where zero is no weight

First we need a

4.40 Lemma. *Let $0\notin\Omega$. Then there is a weight $\lambda\in\Omega$, such that for every root holds* $\left|\frac{2\langle\lambda,\alpha\rangle}{\|\alpha\|^2}\right|\leq 1$.

Proof. Let λ be a weight and α a root such that $\frac{2\langle\lambda,\alpha\rangle}{\|\alpha\|^2}=:k\geq 2$. If k is even we have that $0\neq\lambda-\frac{k}{2}\alpha\in\Omega$. But for this weight holds $\frac{2\langle\lambda-\frac{k}{2}\alpha,\alpha\rangle}{\|\alpha\|^2}=k-k=0$. If k is odd we have that $0\neq\lambda-\frac{k-1}{2}\alpha\in\Omega$ and $\frac{2\langle\lambda-\frac{k-1}{2}\alpha,\alpha\rangle}{\|\alpha\|^2}=1$. □

4.41 Proposition. *Let $\mathfrak{g} \subset \mathfrak{so}(N, \mathbb{C})$ be an irreducible representation of real type of a complex simple Lie algebra different from $\mathfrak{sl}(2, \mathbb{C})$, with $0 \notin \Omega$ and satisfying (SII). Then $\left|\frac{2\langle\Lambda,\beta\rangle}{\|\beta\|^2}\right| \leq 3$ for all roots $\beta \in \Delta$.*

Proof. Let α be in Δ with the property (SII), i.e. $\Omega_\alpha \subset \{\Lambda - \alpha + \beta \mid \beta \in \Delta_0\} \cup \{-\Lambda + \beta \mid \beta \in \Delta_0\}$.
By the previous lemma there is a $\lambda \in \Omega$ such that $\left|\frac{2\langle\lambda,\beta\rangle}{\|\beta\|^2}\right| \leq 1$ for all roots $\beta \in \Delta$. Applying the Weyl group one can choose λ such that $\langle\lambda, \alpha\rangle < 0$.
$\langle\lambda, \alpha\rangle < 0$ implies $\lambda \in \Omega_\alpha$. Hence (SII) gives $\lambda = \Lambda - \alpha - \gamma$ or $\lambda = -\Lambda + \gamma$ with $\gamma \in \Delta_0$. The second case gives for every $\beta \in \Delta$

$$\left|\frac{2\langle\Lambda,\beta\rangle}{\|\beta\|^2}\right| \leq \left|\frac{2\langle\lambda,\beta\rangle}{\|\beta\|^2}\right| + \left|\frac{2\langle\gamma,\beta\rangle}{\|\beta\|^2}\right| \leq 3,$$

because we have excluded G_2.
Thus we have to consider the first case $\Lambda = \lambda + \alpha + \gamma$ with $\gamma \in \Delta_0$ and have to verify that

$$\left|\frac{2\langle\Lambda,\beta\rangle}{\|\beta\|^2}\right| = \left|\frac{2\langle\lambda,\beta\rangle}{\|\beta\|^2} + \frac{2\langle\alpha,\beta\rangle}{\|\beta\|^2} + \frac{2\langle\gamma,\beta\rangle}{\|\beta\|^2}\right| \leq 3. \tag{4.16}$$

for all roots $\beta \in \Delta$.
For $\beta = \pm\alpha$ this is satisfied:

$$\frac{2\langle\Lambda,\alpha\rangle}{\|\alpha\|^2} = \pm\frac{2\langle\lambda,\alpha\rangle}{\|\alpha\|^2} \pm 2 + \frac{2\langle\gamma,\alpha\rangle}{\|\alpha\|^2} = \mp 1 \pm 2 + \frac{2\langle\gamma,\alpha\rangle}{\|\alpha\|^2} \leq 3.$$

Now we have to show (4.16) for all $\beta \in \Delta$ with $\beta \not\sim \alpha$. For this we consider three cases.

Case 1: All roots have the same length. This implies $\left|\frac{2\langle\gamma,\beta\rangle}{\|\beta\|^2}\right| \leq 1$ for all roots which are not proportional to each other. Thus we have (4.16) for all $\beta \not\sim \gamma$:

$$\left|\frac{2\langle\Lambda,\beta\rangle}{\|\beta\|^2}\right| \leq \left|\frac{2\langle\lambda,\beta\rangle}{\|\beta\|^2}\right| + \left|\frac{2\langle\alpha,\beta\rangle}{\|\beta\|^2}\right| + \left|\frac{2\langle\gamma,\beta\rangle}{\|\beta\|^2}\right| \leq 3.$$

For $\beta = \pm\gamma$ we have

$$\frac{2\langle\Lambda,\gamma\rangle}{\|\gamma\|^2} = \frac{2\langle\lambda,\gamma\rangle}{\|\gamma\|^2} + \frac{2\langle\alpha,\gamma\rangle}{\|\gamma\|^2} + 2.$$

This has absolute value ≥ 4 only if $\langle\lambda, \gamma\rangle > 0$ and $\langle\alpha, \gamma\rangle > 0$. This implies that $\alpha - \gamma$ is a root. But for this one holds $\frac{2\langle\lambda,\gamma-\alpha\rangle}{\|\gamma-\alpha\|^2} = \frac{2\langle\lambda,\gamma\rangle}{\|\gamma-\alpha\|^2} - \frac{2\langle\lambda,\alpha\rangle}{\|\gamma-\alpha\|^2} = 2$ since all roots have the same length. This is a contradiction to the choice of λ.

Case 2: There are long and short roots and β is a long root. This implies again $\left|\frac{2\langle\gamma,\beta\rangle}{\|\beta\|^2}\right| \leq 1$ for all β which are not proportional to γ. This implies (4.16) in this case.

For $\beta = \pm\gamma$ we argue as above, recalling that $\gamma - \alpha$ and α have to be short roots in this case. Hence $\frac{2\langle\lambda,\gamma-\alpha\rangle}{\|\gamma-\alpha\|^2} = \frac{2\langle\lambda,\gamma\rangle}{\|\gamma-\alpha\|^2} - \frac{2\langle\lambda,\alpha\rangle}{\|\gamma-\alpha\|^2} \geq \frac{2\langle\Lambda,\gamma\rangle}{\|\gamma\|^2} - \frac{2\langle\lambda,\alpha\rangle}{\|\alpha\|^2} \geq 2$ which is a contradiction.

Case 3: There are long and short roots and β is a short root.

First we consider the case where $\beta = \pm\gamma$. Again (4.16) is not satisfied only if $\langle\lambda,\gamma\rangle$ and $\langle\alpha,\gamma\rangle$ are non zero and have the same sign, lets say $+$.

If α is a short root too, then because of $\langle\alpha,\gamma\rangle \neq 0$ lemma 4.26 gives that $\alpha-\gamma$ is also a short root. Hence $\frac{2\langle\lambda,\gamma-\alpha\rangle}{\|\gamma-\alpha\|^2} = \frac{2\langle\lambda,\gamma\rangle}{\|\gamma-\alpha\|^2} - \frac{2\langle\lambda,\alpha\rangle}{\|\gamma-\alpha\|^2} = \frac{2\langle\lambda,\gamma\rangle}{\|\gamma\|^2} - \frac{2\langle\lambda,\alpha\rangle}{\|\alpha\|^2} = 2$ yields a contradiction.

If α is a long root, then $\gamma-\alpha$ has to be a short one and we get again a contradiction: $\frac{2\langle\lambda,\gamma-\alpha\rangle}{\|\gamma-\alpha\|^2} = \frac{2\langle\lambda,\gamma\rangle}{\|\gamma-\alpha\|^2} - \frac{2\langle\lambda,\alpha\rangle}{\|\gamma-\alpha\|^2} \geq \frac{2\langle\lambda,\gamma\rangle}{\|\gamma\|^2} - \frac{2\langle\lambda,\alpha\rangle}{\|\alpha\|^2} \geq 2$.

Now suppose that $\beta \not\sim \gamma$. Then $\frac{2\langle\Lambda,\beta\rangle}{\|\beta\|^2} = \frac{2\langle\lambda,\beta\rangle}{\|\beta\|^2} + \frac{2\langle\alpha,\beta\rangle}{\|\beta\|^2} + \frac{2\langle\gamma,\beta\rangle}{\|\beta\|^2}$ has absolute value ≥ 4 only if all three right hand side terms have the same sign — lets say they are positive — and at least one of the last two terms has absolute value greater than one, i.e.γ or α is a long root. If α is a long root then $\alpha-\beta$ is a short one and arguing as above gives the contradiction. If α is a short root then $\langle\alpha,\beta\rangle > 0$ implies by lemma 4.26 that $\beta-\alpha$ is a short root. Again we have a contradiction: $\frac{2\langle\lambda,\beta-\alpha\rangle}{\|\beta-\alpha\|^2} = \frac{2\langle\lambda,\beta\rangle}{\|\beta-\alpha\|^2} - \frac{2\langle\lambda,\alpha\rangle}{\|\beta-\alpha\|^2} = \frac{2\langle\lambda,\beta\rangle}{\|\beta\|^2} - \frac{2\langle\lambda,\alpha\rangle}{\|\alpha\|^2} = 2$.

□

4.42 Proposition. *Under the same assumptions as in the previous proposition holds that* $\left|\frac{2\langle\Lambda,\eta\rangle}{\|\eta\|^2}\right| \leq 2$ *for all long roots* η.

Proof. Let Λ and α be the extremal weight and the root from property (SII). We suppose that there is a long root η with

$$\frac{2\langle\Lambda,\eta\rangle}{\|\eta\|^2} = -3 \tag{4.17}$$

and derive a contradiction considering different cases.

Case 1: All roots have the same length. By applying the Weyl group we find an extremal weight Λ' such that $a := \frac{2\langle\Lambda',\alpha\rangle}{\|\alpha\|^2} = -3$.

First we find a root β with

$$\frac{2\langle\alpha,\beta\rangle}{\|\beta\|^2} = 1 \text{ and } \frac{2\langle\Lambda',\beta\rangle}{\|\beta\|^2} \leq -2.$$

This is obvious: We find a β such that $\frac{2\langle\alpha,\beta\rangle}{\|\beta\|^2} = 1$. If $\frac{2\langle\Lambda',\beta\rangle}{\|\beta\|^2} \geq -1$ then we consider the root $\alpha-\beta$. It satisfies $\frac{2\langle\alpha,\alpha-\beta\rangle}{\|\alpha-\beta\|^2} = 1$ and we have

$$\frac{2\langle\Lambda',\alpha-\beta\rangle}{\|\alpha-\beta\|^2} = -3 - \frac{2\langle\Lambda',\beta\rangle}{\|\alpha-\beta\|^2} \leq -2.$$

Hence we have $\Lambda' + k\beta \in \Omega$ for $0 \leq k \leq 2$ and $\Lambda' + k\alpha \in \Omega$ for $0 \leq k \leq 3$. Furthermore

$$\frac{2\langle \Lambda' + l\beta, \alpha\rangle}{\|\alpha\|^2} = -3 - \frac{2\langle \Lambda', \alpha\rangle}{\|\alpha\|^2} = -3 + l.$$

But this gives

$$\Lambda' + k\alpha + l\beta \in \Omega_\alpha \text{ for } 0 \leq k \leq 2, 0 \leq k + l \leq 2.$$

Among others (SII) implies the existence of γ_i and δ_i from Δ_0 for $i = 0, 1, 2$ such that that the following alternatives must hold

$$\begin{array}{rclcrcll}
\Lambda' + \alpha & = & \Lambda + \gamma_0 & \text{or} & \Lambda' & = & -\Lambda + \delta_0 & (4.18) \\
\Lambda' + 3\alpha & = & \Lambda + \gamma_1 & \text{or} & \Lambda' + 2\alpha & = & -\Lambda + \delta_1 & (4.19) \\
\Lambda' + \alpha + 2\beta & = & \Lambda + \gamma_2 & \text{or} & \Lambda' + 2\beta & = & -\Lambda + \delta_2. & (4.20)
\end{array}$$

First we suppose that the first alternative of (4.18) holds, i.e $\Lambda' + \alpha = \Lambda + \gamma_0$. Since $a = -3$ and both Λ and Λ' are extremal we have that $\alpha \neq -\gamma_0$. Hence the first case of (4.19) cannot be true and we have $\Lambda' + 2\alpha = -\Lambda + \delta_1$. We consider now (4.20): The left side of (4.18) gives that $\Lambda' + 2\beta + \alpha = \Lambda + \gamma_0 + 2\beta$. If the left side of (4.20) holds, we would have $\gamma_0 = -\beta$. Hence $\Lambda + \beta \in \Omega$ and on the other hand $\Omega \ni \Lambda' + \alpha = \Lambda - \beta$ which contradicts the extremality of Λ. Thus the right hand side of (4.20) must be satisfied. From $\Lambda' + 2\alpha = -\Lambda + \delta_1$ follows $\Lambda' + 2\beta = -\Lambda + \delta_1 + 2(\beta - \alpha)$ and therefore $\delta_1 = -(\beta - \alpha)$. Again we have $-\Lambda + (\beta - \alpha) \in \Omega$ and $-\Lambda - (\beta - \alpha) \in \Omega$ which contradicts the extremality of Λ.

If one starts with the right hand side of (4.18) we can proceed analogously and get a contradiction in the case where all roots have the same length.

Case 2. The roots have different length and α is a short root. On one hand we find a short root β which is orthogonal to α and $\alpha + \beta$ is a long root, and on the other we can find an extremal weight Λ' such that

$$\frac{2\langle \Lambda', \alpha + \beta\rangle}{\|\alpha + \beta\|^2} = -3.$$

Since $\alpha \perp \beta$ we have

$$-3 = \frac{2(\langle \Lambda', \alpha\rangle + \langle \Lambda', \beta\rangle)}{\|\alpha\|^2 + \|\beta\|^2} = \frac{1}{2}\left(\frac{2\langle \Lambda', \alpha\rangle}{\|\alpha\|^2} + \frac{2\langle \Lambda', \beta\rangle}{\|\beta\|^2}\right).$$

Because of the previous proposition we get

$$\frac{2(\langle \Lambda', \alpha\rangle}{\|\alpha\|^2} = \frac{2(\langle \Lambda', \beta\rangle}{\|\beta\|^2} = -3.$$

Hence $\Lambda' + k\alpha + l\beta \in \Omega$ for $0 \leq k, l \leq 3$ and therefore $\Lambda' + k\alpha + l\beta \in \Omega_\alpha$ for $0 \leq k \leq 2$ and $0 \leq l \leq 3$. (SII) implies the following alternatives

$$\begin{array}{rclcrcll}
\Lambda' + \alpha & = & \Lambda + \gamma_0 & \text{or} & \Lambda' & = & -\Lambda + \delta_0 & (4.21) \\
\Lambda' + \alpha + 3\beta & = & \Lambda + \gamma_1 & \text{or} & \Lambda' + 3\beta & = & -\Lambda + \delta_1 & (4.22) \\
\Lambda' + 2\alpha + 3\beta & = & \Lambda + \gamma_2 & \text{or} & \Lambda' + \alpha + 3\beta & = & -\Lambda + \delta_2 & (4.23) \\
\Lambda' + 3\alpha + 2\beta & = & \Lambda + \gamma_3 & \text{or} & \Lambda' + 2(\alpha + \beta) & = & -\Lambda + \delta_3 & (4.24) \\
\Lambda' + 3\alpha + 3\beta & = & \Lambda + \gamma_4 & \text{or} & \Lambda' + 2\alpha + 3\beta & = & -\Lambda + \delta_4. & (4.25)
\end{array}$$

If the left hand side of the first alternative is valid then the left hand sides of the remaining four cannot be satisfied: For (4.22) we would have $3\beta = \gamma_1 - \gamma_0$ which is not possible. (4.23) would imply $3\beta + \alpha = \gamma_2 - \gamma_0$ which is by lemma 4.29 a contradiction since $\alpha \neq -\beta$ and $\gamma_0 \neq -\alpha$. (4.24) would imply $2(\alpha+\beta) = \gamma_3 - \gamma_0$. Since $\alpha + \beta$ is a long root this would give $\gamma_0 = -\alpha + \beta$ and $\gamma_3 = \alpha + \beta$ which is a contradiction to the extremality of Λ. (4.25) would give $2\alpha + 3\beta = \gamma_4 - \gamma_0$ which also is not possible.

Thus for the last four equations the right hand side must hold. Taking everything together we would get $\alpha = \delta_2 - \delta_1 = \delta_4 - \delta_2$ and $\beta = \delta_4 - \delta_3$. This gives $2\alpha = \delta_4 - \delta_1$ and thus

$$\frac{2\langle \delta_4, \alpha \rangle}{\|\alpha\|^2} - \frac{2\langle \delta_1, \alpha \rangle}{\|\alpha\|^2} = \frac{4\|\alpha\|^2}{\|\alpha\|^2} = 4.$$

The extremality of Λ prevents that $\alpha = \delta_4 = -\delta_1$. Hence δ_1 and δ_4 are long roots, in particular

$$\frac{2\langle \delta_4, \alpha \rangle}{\|\alpha\|^2} = -\frac{2\langle \delta_1, \alpha \rangle}{\|\alpha\|^2} = 2.$$

For β again $\beta = \delta_4 = -\delta_3$ cannot hold by the extremality of Λ and we have

$$0 = \frac{2\langle \beta, \alpha \rangle}{\|\alpha\|^2} = \frac{2\langle \delta_4, \alpha \rangle}{\|\alpha\|^2} - \frac{2\langle \delta_3, \alpha \rangle}{\|\alpha\|^2} = 2 - \frac{2\langle \delta_3, \alpha \rangle}{\|\alpha\|^2}$$

which forces δ_3 to be a long root too. Now we have a contradiction because the short root β is the sum of two long roots. This is impossible.

If we start with the right hand side of the first alternative one proceeds analogously.

Case 3. The roots have different length and α is a long root. In this case we find an extremal weight Λ' such that $\frac{2\langle \Lambda', \alpha \rangle}{\|\alpha\|^2} = -3$. Now we can write $\alpha = \alpha_1 + \alpha_2$ with $\alpha_1 \perp \alpha_2$ two short roots. As above we get

$$\frac{2(\langle \Lambda', \alpha_1 \rangle}{\|\alpha_1\|^2} = \frac{2(\langle \Lambda', \alpha_2 \rangle}{\|\alpha_2\|^2} = -3. \tag{4.26}$$

Again this implies $\Lambda'+k\alpha+l\beta \in \Omega$ for $0 \le k,l \le 3$ and therefore $\Lambda'+k\alpha+l\beta \in \Omega_\alpha$ for $0 \le k,l \le 2$. Now (SII) implies the existence of γ_i and δ_i from Δ_0 for $i = 0,\dots,8$ such that that the following alternatives must hold

$$
\begin{array}{rclcrcll}
 & (L) & & & & (R) & & \\
\Lambda' + \alpha_1 + \alpha_2 & = & \Lambda + \gamma_0 & \text{or} & \Lambda' & = & -\Lambda + \delta_0 & (4.27) \\
\Lambda' + 2\alpha_1 + \alpha_2 & = & \Lambda + \gamma_1 & \text{or} & \Lambda' + \alpha_1 & = & -\Lambda + \delta_1 & (4.28) \\
\Lambda' + 3\alpha_1 + \alpha_2 & = & \Lambda + \gamma_2 & \text{or} & \Lambda' + 2\alpha_1 & = & -\Lambda + \delta_2 & (4.29) \\
\Lambda' + \alpha_1 + 2\alpha_2 & = & \Lambda + \gamma_3 & \text{or} & \Lambda' + \alpha_2 & = & -\Lambda + \delta_3 & (4.30) \\
\Lambda' + \alpha_1 + 3\alpha_2 & = & \Lambda + \gamma_4 & \text{or} & \Lambda' + 2\alpha_2 & = & -\Lambda + \delta_4 & (4.31) \\
\Lambda' + 2\alpha_1 + 2\alpha_2 & = & \Lambda + \gamma_5 & \text{or} & \Lambda' + \alpha_1 + \alpha_2 & = & -\Lambda + \delta_5 & (4.32) \\
\Lambda' + 2\alpha_1 + 3\alpha_2 & = & \Lambda + \gamma_6 & \text{or} & \Lambda' + \alpha_1 + 2\alpha_2 & = & -\Lambda + \delta_6 & (4.33) \\
\Lambda' + 3\alpha_1 + 2\alpha_2 & = & \Lambda + \gamma_7 & \text{or} & \Lambda' + 2\alpha_1 + \alpha_2 & = & -\Lambda + \delta_7 & (4.34) \\
\Lambda' + 3\alpha_1 + 3\alpha_2 & = & \Lambda + \gamma_8 & \text{or} & \Lambda' + 2\alpha_1 + 2\alpha_2 & = & -\Lambda + \delta_8. & (4.35)
\end{array}
$$

In the following we denote the left hand side formulas with an .L and the right hand side formulas with an .R. Again we suppose that (4.27.L) is satisfied, i.e. $\Lambda' + \alpha_1 + \alpha_2 = \Lambda + \gamma_0$. Then (4.26) and the extremality of Λ implies that γ_0 does not equal to α_i.

Now (4.35.L) would imply that $2(\alpha_1 + \alpha_2) = 2\alpha = \gamma_8 - \gamma_0$. Since α is a long root this is not possible and we have (4.35.R), i.e. $\Lambda' + 2\alpha_1 + 2\alpha_2 = -\Lambda + \delta_8$.

Thinking for a moment gives that (4.34.L) implies $\gamma_0 = -\alpha_1$ and (4.33.L) implies $\gamma_0 = -\alpha_2$. On the other hand (4.31.L) implies $\gamma_0 \neq -\alpha_1$ and (4.29.L) implies $\gamma_0 \neq -\alpha_2$. Hence (4.34.L) entails (4.33.R) and (4.31.R), as well as (4.33.L) entails (4.34.R) and (4.29.R).

Now we suppose that (4.34.L) is satisfied. Then we have (4.35.R), (4.33.R) and (4.31.R), i.e.

$$\alpha_2 = \delta_8 - \delta_7 \text{ and } 2\alpha_1 = \delta_8 - \delta_4.$$

Again because of the extremality of Λ these roots are not proportional. It implies $2\alpha_1 - \alpha_2 = \delta_7 - \delta_4$. Now $\alpha_1 \perp \alpha_2$ and $\delta_7 \neq \alpha_1, \neq \alpha_1 - \alpha_2$ (Extremality of Λ) gives a contradiction.

In the same way we argue supposing that (4.33.L) holds.

Hence we have shown that neither (4.33.L) nor (4.34.L) can be satisfied. Thus we have (4.33.R) and (4.34.R). These together with (4.35.L) are no contradiction, but if one supposes one of the remaining (4.28.R), (4.30.R) or (4.32.R) we get a contradiction. Hence (4.28.L), (4.30.L) and (4.32.L) must be valid. But from these together with (4.27.L) we derive as above a contradiction.

If we start with the right hand side of the first alternative one proceeds analogously.

All in all we have shown, that the assumption of a long root with (4.17) leads to a contradiction. □

Now we are in a position that we can use results of [Sch99] explicitly. First we will cite them.

4.43 Proposition. *[Sch99] Let $\mathfrak{g} \subset \mathfrak{so}(N,\mathbb{C})$ be an irreducible representation of real type of a complex simple Lie algebra different from $\mathfrak{sl}(2,\mathbb{C})$. Then it holds:*

1. *If there is an extremal spanning $(\Lambda_1, \Lambda_2, \alpha)$ triple then there is no weight λ for which exists a pair of orthogonal long roots η_1 and η_2 such that $\left|\frac{2\langle\lambda,\eta_i\rangle}{\|\eta_i\|^2}\right| = 2$.*
2. *If furthermore all roots have the same length, then there is no weight λ for which exists a triple of orthogonal roots $\eta_1 \perp \eta_2 \perp \eta_3 \perp \eta_1$ such that $\left|\frac{2\langle\lambda,\eta_1\rangle}{\|\eta_1\|^2}\right| = 2$ and $\left|\frac{2\langle\lambda,\eta_2\rangle}{\|\eta_2\|^2}\right| = \left|\frac{2\langle\lambda,\eta_3\rangle}{\|\eta_3\|^2}\right| = 1$.*

We will now show that existence of such a pair or triple of roots implies that (SII) defines an extremal spanning pair.

4.44 Proposition. *Let $\mathfrak{g} \subset \mathfrak{so}(N,\mathbb{C})$ be an irreducible representation of real type of a complex simple Lie algebra different from $\mathfrak{sl}(2,\mathbb{C})$, with $0 \notin \Omega$ and satisfying (SII). Then it holds: If there is a pair of orthogonal long roots η_1 and η_2 such that $\left|\frac{2\langle\Lambda,\eta_i\rangle}{\|\eta_i\|^2}\right| = 2$ for the extremal weight Λ from the property (SII), then $\Lambda - \alpha$ is an extremal weight, i.e. (SII) defines an extremal spanning triple.*

Proof. Again we argue indirectly considering three different cases for the root α from the property (SII)

Case 1: All roots have the same length or α is a long root. Again by applying the Weyl group the indirect assumption implies that there is an extremal weight Λ' and a root long β orthogonal to α such that $\frac{2\langle\Lambda',\alpha\rangle}{\|\alpha\|^2} = \frac{2\langle\Lambda',\beta\rangle}{\|\beta\|^2} = -2$.

This gives that $\Lambda' + k\alpha + l\beta \in \Omega$ for $0 \leq k, l \leq 2$ and hence $\Lambda' + k\alpha + l\beta \in \Omega_\alpha$ for $0 \leq k \leq 1, 0 \leq l \leq 2$.

Among others (SII) implies the existence of γ_i and δ_i from Δ_0 for $i = 0, \ldots, 3$

such that that the following alternatives must hold

$$\begin{array}{rclcrcll} & (L) & & & & (R) & & \\ \Lambda' + \alpha & = & \Lambda + \gamma_0 & \text{or} & \Lambda' & = & -\Lambda + \delta_0 & (4.36) \\ \Lambda' + 2\alpha & = & \Lambda + \gamma_1 & \text{or} & \Lambda' + \alpha & = & -\Lambda + \delta_1 & (4.37) \\ \Lambda' + \alpha + 2\beta & = & \Lambda + \gamma_2 & \text{or} & \Lambda' + 2\beta & = & -\Lambda + \delta_2 & (4.38) \\ \Lambda' +_1 + 2\alpha + 2\beta & = & \Lambda + \gamma_3 & \text{or} & \Lambda' + \alpha + 2\beta & = & -\Lambda + \delta_3. & (4.39) \end{array}$$

Supposing again (4.36.L) we conclude that (4.38.L) and (4.39.L) cannot hold because β is long and the extremality of Λ. Hence (4.38.R) and (4.39.R) must be satisfied. Again the extremality of Λ prevents that (4.37.R) can be valid. Hence we have (4.37.L).

Now (4.36.L) gives that

$$\frac{2\langle \Lambda, \alpha \rangle}{||\alpha||^2} = \frac{2\langle \Lambda', \alpha \rangle}{||\alpha||^2} + 2 - \frac{2\langle \gamma_0, \alpha \rangle}{||\alpha||^2} = -\frac{2\langle \gamma_0, \alpha \rangle}{||\alpha||^2}$$

by assumption.

On the other hand (4.36.L) together with (4.37.L) and (4.38.R) and (4.39.R) implies $\alpha = \gamma_1 - \gamma_0 = \delta_3 - \delta_2$. We note that γ_0 cannot be equal to 0 and γ_1 not equal to α.

If $\gamma_0 = -\alpha$ and $\gamma_1 = 0$ then $\Lambda = \Lambda' + 2\alpha$. Then (4.38.R) and (4.39.R) imply

$$\begin{array}{rclcl} \langle \delta_2, \alpha \rangle & = & 2\langle \Lambda', \alpha \rangle + 2||\alpha||^2 & = & 0 \text{ and} \\ \langle \delta_3, \alpha \rangle & = & 2\langle \Lambda', \alpha \rangle + 3||\alpha||^2 & = & ||\alpha||^2. \end{array}$$

Since α is long this entails $\delta_2 = 0$ and $\delta_3 = \alpha$. Taking now (4.36.L) and (4.38.R) together we get $\Lambda = \alpha - \beta$. But this forces $0 \in \Omega$ which was excluded.

Thus we have $\alpha = \gamma_1 - \gamma_0$ with non-proportionality. But this implies, since α is long, that $\frac{2\langle \gamma_0, \alpha \rangle}{||\alpha||^2} = -1$ and hence $\frac{2\langle \Lambda, \alpha \rangle}{||\alpha||^2} = 1$. But this means that $\Lambda - \alpha$ is an extremal weight.

Case 2: There are roots with different length and α is a short root. By assumption there is a short root γ such that $\gamma \perp \alpha$ and $\eta := \alpha + \gamma$ is a long root and an extremal weight Λ' and a long root β such that $\frac{2\langle \Lambda', \eta \rangle}{||\eta||^2} = \frac{2\langle \Lambda', \beta \rangle}{||\beta||^2} = -2$. Analogously to the previous theorem the orthogonality of α and γ gives

$$-2 = \frac{1}{2} \left(\frac{2\langle \Lambda', \alpha \rangle}{||\alpha||^2} + \frac{2\langle \Lambda', \gamma \rangle}{||\gamma||^2} \right).$$

Hence we have to consider three cases:

(a) $\frac{2\langle\Lambda',\alpha\rangle}{\|\alpha\|^2} = \frac{2\langle\Lambda',\gamma\rangle}{\|\gamma\|^2} - 2$,

(b) $\frac{2\langle\Lambda',\alpha\rangle}{\|\alpha\|^2} = -3$ and $\frac{2\langle\Lambda',\gamma\rangle}{\|\gamma\|^2} - 1$,

(c) $\frac{2\langle\Lambda',\alpha\rangle}{\|\alpha\|^2} = -1$ and $\frac{2\langle\Lambda',\gamma\rangle}{\|\gamma\|^2} - 3$.

Then an easy calculation shows that $\langle\alpha,\beta\rangle = \langle\gamma,\beta\rangle = 0$ in each case.

We shall consider the case (a),(b) and (c) separately.

Case (a): Here we can proceed completely analogously to the first case 1. We have that $\Lambda' + k\alpha + l\beta \in \Omega_\alpha$ for $0 \le k \le 1, 0 \le l \le 2$ leading to the same set of equations (4.36) — (4.39) and the same implications since β is long again. The proportional case is excluded as above and we get that $\alpha = \gamma_1 - \gamma_0$ non proportional. At least one has to be a short root and $\langle\gamma_0,\alpha\rangle < 0$ and $\langle\gamma_1,\alpha\rangle > 0$. On the other hand we have $\frac{2\langle\Lambda,\alpha\rangle}{\|\alpha\|^2} = -\frac{2\langle\gamma_0,\alpha\rangle}{\|\alpha\|^2}$ and $\frac{2\langle\Lambda,\alpha\rangle}{\|\alpha\|^2} = -\frac{2\langle\gamma_1,\alpha\rangle}{\|\alpha\|^2} + 2$ by (4.36.L) and (4.37.L). But this implies that both are short and $\frac{2\langle\Lambda,\alpha\rangle}{\|\alpha\|^2} = 1$ which is the proposition.

Case (b): $\frac{2\langle\Lambda',\alpha\rangle}{\|\alpha\|^2} = -3$ implies that $\Lambda' + k\alpha + l\beta \in \Omega_\alpha$ for $0 \le k,l \le 2$. (SII) then implies

$$
\begin{array}{rclcrcll}
 & (L) & & & & (R) & & \\
\Lambda' + \alpha & = & \Lambda + \gamma_0 & \text{or} & \Lambda' & = & -\Lambda + \delta_0 & (4.40) \\
\Lambda' + 2\alpha & = & \Lambda + \gamma_1 & \text{or} & \Lambda' + \alpha & = & -\Lambda + \delta_1 & (4.41) \\
\Lambda' + 3\alpha & = & \Lambda + \gamma_2 & \text{or} & \Lambda' + 2\alpha & = & -\Lambda + \delta_2 & (4.42) \\
\Lambda' + 2\alpha + 2\beta & = & \Lambda + \gamma_3 & \text{or} & \Lambda' + \alpha + 2\beta & = & -\Lambda + \delta_3 & (4.43) \\
\Lambda' + 3\alpha + 2\beta & = & \Lambda + \gamma_4 & \text{or} & \Lambda' + 2\alpha + 2\beta & = & -\Lambda + \delta_3. & (4.44)
\end{array}
$$

Supposing (4.40.L) excludes (4.43.L) and (4.44.L) because β is long. Hence (4.43.R) and (4.44.R) are valid and exclude (4.41.R) and (4.42.L). Hence (4.41.L) and (4.42.L) are satisfied. This gives $\alpha = \gamma_2 - \gamma_1 = \gamma_1 - \gamma_0$ with γ_0 different from 0 and $-\alpha$, γ_1 different from 0 and α and γ_2 different from $\pm\alpha$. Hence $\alpha + \pm\delta_1$ is a long root with $\alpha \perp \delta_1$. But this gives $\frac{2\langle\Lambda,\alpha\rangle}{\|\alpha\|^2} = \frac{2\langle\Lambda',\alpha\rangle}{\|\alpha\|^2} + 4 = 1$, i.e.$\Lambda - \alpha$ is an extremal weight.

Case (c): Here we have that $\Lambda' + k\gamma + l\beta \in \Omega_\alpha$ for $0 \le k \le 3$ and $0 \le l \le 2$ since $\frac{2\langle\Lambda' + k\gamma + l\beta,\alpha\rangle}{\|\alpha\|^2} = -1$. The equations implied by (SII) lead easily to a contradiction:

$$\begin{array}{rclcrcll} & (L) & & & & (R) & & \\ \Lambda' + \alpha & = & \Lambda + \gamma_0 & \text{or} & \Lambda' & = & -\Lambda + \delta_0 & (4.45) \\ \Lambda' + \alpha + 3\gamma & = & \Lambda + \gamma_1 & \text{or} & \Lambda' + 3\gamma & = & -\Lambda + \delta_1 & (4.46) \\ \Lambda' + \alpha + 2\beta + 3\gamma & = & \Lambda + \gamma_2 & \text{or} & \Lambda' + 2\beta + 3\gamma & = & -\Lambda + \delta_2. & (4.47) \end{array}$$

Supposing (4.45.L) excludes (4.46.L) and (4.47.L). Hence (4.46.R) and (4.47.R) are valid but contradict to each other because β is long.

□

4.45 Proposition. *Let $\mathfrak{g} \subset \mathfrak{so}(N, \mathbb{C})$ be an irreducible representation of real type of a complex simple Lie algebra different from $\mathfrak{sl}(2, \mathbb{C})$, with $0 \notin \Omega$ and satisfying (SII). If furthermore all roots have the same length, and if there is a triple of orthogonal roots $\eta_1 \perp \eta_2 \perp \eta_3 \perp \eta_1$ such that $\left|\frac{2\langle\Lambda,\eta_1\rangle}{\|\eta_1\|^2}\right| = 2$ and $\left|\frac{2\langle\Lambda,\eta_2\rangle}{\|\eta_2\|^2}\right| = \left|\frac{2\langle\Lambda,\eta_3\rangle}{\|\eta_3\|^2}\right| = 1$ then one of the cases holds*

1. *$\Lambda - \alpha$ is an extremal weight, i.e. (SII) defines an extremal spanning triple, or*
2. *$\Lambda = \alpha + \frac{1}{2}(\beta + \gamma)$ with roots $\alpha \perp \beta \perp \gamma \perp \alpha$.*

Proof. Let α be the root determined by (SII). The assumption implies that there is an extremal weight Λ' and roots β and γ such that

$$\frac{2\langle\Lambda', \alpha\rangle}{\|\alpha\|^2} = -2 \text{ and } \frac{2\langle\Lambda, \beta\rangle}{\|\beta\|^2} = \frac{2\langle\Lambda, \gamma\rangle}{\|\gamma\|^2} = -1.$$

Then $\Lambda' + k\alpha + l\beta + m\gamma \in \Omega$ for $k, l, m = 0, 1$. Hence (SII) implies again

$$\begin{array}{rclcrcll} & (L) & & & & (R) & & \\ \Lambda' + \alpha & = & \Lambda + \gamma_0 & \text{or} & \Lambda' & = & -\Lambda + \delta_0 & (4.48) \\ \Lambda' + 2\alpha & = & \Lambda + \gamma_1 & \text{or} & \Lambda' + \alpha & = & -\Lambda + \delta_1 & (4.49) \\ \Lambda' + \alpha + \beta & = & \Lambda + \gamma_2 & \text{or} & \Lambda' + \beta & = & -\Lambda + \delta_2 & (4.50) \\ \Lambda' + 2\alpha + \beta & = & \Lambda + \gamma_3 & \text{or} & \Lambda' + \alpha + \beta & = & -\Lambda + \delta_3 & (4.51) \\ \Lambda' + \alpha + \gamma & = & \Lambda + \gamma_4 & \text{or} & \Lambda' + \gamma & = & -\Lambda + \delta_4 & (4.52) \\ \Lambda' + 2\alpha + \gamma & = & \Lambda + \gamma_5 & \text{or} & \Lambda' + \alpha + \gamma & = & -\Lambda + \delta_5 & (4.53) \\ \Lambda' + \alpha + \beta + \gamma & = & \Lambda + \gamma_6 & \text{or} & \Lambda' + \beta + \gamma & = & -\Lambda + \delta_6 & (4.54) \\ \Lambda' + 2\alpha + \beta + \gamma & = & \Lambda + \gamma_7 & \text{or} & \Lambda' + \alpha + \beta + \gamma & = & -\Lambda + \delta_7. & (4.55) \end{array}$$

Supposing (4.48.L) excludes (4.55.R) because of the orthogonality of the roots. Thus it must hold (4.55.R). Now we consider two cases:

Case 1: $\langle\gamma_0,\beta\rangle = \langle\gamma_0,\gamma\rangle = 0$. This excludes (4.50.L), (4.52.L) and (4.54.L) and implies therefore (4.50.R), (4.52.R) and (4.54.R). The latter together with (4.55.R) gives $\alpha = \delta_7 - \delta_6$.

Since $\delta_7 \neq 0$ this implies $\langle\alpha,\delta_7\rangle > 0$. On the other hand (4.55.R) and the assumption gives that $\frac{2\langle\Lambda,\alpha\rangle}{\|\alpha\|^2} = \frac{2\langle\delta_7,\alpha\rangle}{\|\alpha\|^2} > 0$. If $\alpha \neq \delta_7$ we are done.

But $\delta_7 = \alpha$ implies $\Lambda' + \beta + \gamma = -\Lambda = -\Lambda' - \alpha - \gamma_0$ and hence $-2 = 2 - 2 - \frac{2\langle\alpha,\gamma_0\rangle}{\|\alpha\|^2}$, i.e. $\gamma_0 = -\alpha$. Taking everything together we get $2\Lambda = 2\alpha - (\beta+\gamma)$.

Case 2: $\langle\gamma_0,\beta\rangle$ *or* $\langle\gamma_0,\gamma\rangle$ *not equal to zero.* This implies $\gamma_0 \neq \pm\alpha$ and thus $\frac{2\langle\Lambda,\alpha\rangle}{\|\alpha\|^2} = -\frac{2\langle\gamma_0,\alpha\rangle}{\|\alpha\|^2} = \pm 1$ or zero. Now (4.49.L) would imply $\alpha = \gamma_1 - \gamma_0$, i.e.$\langle\alpha,\gamma_0\rangle < 0$. This would be the proposition.

Hence we suppose (4.49.R). This together with the starting point (4.48.L) gives

$$\begin{aligned}\Lambda &= \frac{1}{2}(\delta_1 - \gamma_0) \quad \text{and}\\ \Lambda' &= -\alpha + \frac{1}{2}(\delta_1 + \gamma_0).\end{aligned}$$

The second equation implies using the assumption that $\langle\alpha,\delta_1+\gamma_0\rangle = 0$. For the length of both extremal weights holds then

$$\begin{aligned}\|\Lambda\|^2 &= \frac{1}{4}\left(\|\delta_1\|^2 + \|\gamma_0\|^2 - 2\langle\delta_1,\gamma_0\rangle\right)\\ \|\Lambda'\|^2 &= \|\alpha\|^2 - \underbrace{\langle\alpha,\delta_1+\gamma_0\rangle}_{=0} + \frac{1}{4}\left(\|\delta_1\|^2 + \|\gamma_0\|^2 + 2\langle\delta_1,\gamma_0\rangle\right).\end{aligned}$$

This gives $0 = \|\alpha\|^2 + \langle\delta_1,\gamma_0\rangle$. Since all roots have the same length this implies $\delta_1 = -\gamma_0$. Hence Λ is a root. But this was excluded.

□

Now using the proposition 4.43 of Schwachhöfer we get a corollary.

4.46 Corollary. *Let $\mathfrak{g} \subset \mathfrak{so}(N,\mathbb{C})$ be an irreducible representation of real type of a complex simple Lie algebra different from $\mathfrak{sl}(2,\mathbb{C})$, with $0 \notin \Omega$ and satisfying (SII). Then it holds:*

1. *There is no a pair of orthogonal long roots η_1 and η_2 such that $\left|\frac{2\langle\Lambda,\eta_i\rangle}{\|\eta_i\|^2}\right| = 2$ for the extremal weight Λ from the property (SII).*

2. *If furthermore all roots have the same length, and if there is a triple of orthogonal roots $\eta_1 \bot \eta_2 \bot \eta_3 \bot \eta_1$ such that $\left|\frac{2\langle\Lambda,\eta_1\rangle}{\|\eta_1\|^2}\right| = 2$ and $\left|\frac{2\langle\Lambda,\eta_2\rangle}{\|\eta_2\|^2}\right| = \left|\frac{2\langle\Lambda,\eta_3\rangle}{\|\eta_3\|^2}\right| = 1$ then $\Lambda = \alpha + \frac{1}{2}(\beta+\gamma)$ with roots $\alpha\bot\beta\bot\gamma\bot\alpha$.*

Before we apply this corollary we have to deal with the remaining exception in the second point.

4.47 Lemma. *If the representation of a simple Lie algebra with roots of the same length has an extremal weight* Λ *such that* $\Lambda = \alpha + \frac{1}{2}(\beta + \gamma)$ *with roots* $\alpha \perp \beta \perp \gamma \perp \alpha$. *Then it holds*

1. *There is no root* δ *such that* $\langle \delta, \beta \rangle = 0$, $\langle \delta, \gamma \rangle \neq 0$ *and* $\delta \not\perp \gamma$.

2. *The root system is* D_n *and the representation has one of the following highest weights:* ω_3 *for arbitrary* n, $\omega_1 + \omega_3$ *or* $\omega_1 + \omega_4$ *for* $n = 4$ *and* ω_2 *for* $n = 3$.

Proof. The first point is easy to see: If there is such a δ then we have

$$\frac{2\langle \Lambda, \delta \rangle}{\|\delta\|^2} = \frac{2\langle \alpha, \delta \rangle}{\|\delta\|^2} + \frac{1}{2}\frac{2\langle \gamma, \delta \rangle}{\|\delta\|^2} = \frac{2\langle \alpha, \delta \rangle}{\|\delta\|^2} \pm \frac{1}{2} \notin \mathbb{Z}.$$

This is a contradiction.
Now we consider the different root systems with roots of constant length.

A_n: Here the assumption means that $\Lambda = e_i - e_j + \frac{1}{2}(e_p - e_q + e_r - e_s)$ with all indices different from each other. But then $\frac{2\langle \Lambda, e_i - e_p \rangle}{\|e_i - e_p\|^2}$ is not an integer.

D_n: If $\alpha = e_i \pm e_j$, $\beta = e_p \pm e_q$ and $\gamma = e_r \pm e_s$ with all indices different we get the same contradiction as in the A_n case. Thus we are left with two cases.

The first is $\beta + \gamma = e_p + e_q + e_p - e_q = 2e_p$ and hence $\Lambda = e_i \pm e_j + e_p$. This leads to $\Lambda = \omega_3$ or for $n = 3$ to $\Lambda = \omega_2$.

The second is $\alpha = e_i + e_j$, $\beta = e_i - e_j$ and $\gamma = e_p \pm e_q$. For $n > 4$ we found a root $e_p + e_s$ which leads to a contradiction by applying the first point. For $n = 4$ we have $\Lambda = \frac{3}{2}e_i + \frac{1}{2}(e_j + e_p \pm e_q)$. But this yields the remaining representations.

E_6: E_6 has two different types of roots:

$$e_i \pm e_j \quad \text{and} \quad \frac{1}{2}\big(e_8 - e_7 - e_6 \underbrace{\pm e_5 \pm e_4 \pm e_3 \pm e_2 \pm e_1}_{\text{even number of minus signs}}\big).$$

The only possibility for β and γ for which the first point yields no contradiction is $\beta = e_i + e_j$ and $\gamma = e_i - e_j$. Hence $\Lambda = \alpha + e_i$. $\alpha \perp \beta$ and $\alpha \perp \gamma$ implies $\alpha = e_p + e_q$. But then $\langle \Lambda, \frac{1}{2}(\ldots) \rangle \notin \mathbb{Z}$.

Proceeding analogously for E_7 and E_8 we prove the second assertion. □

Now using all these properties we can find the representations without weight zero and satisfying (SII).

4.48 Proposition. *Let $\mathfrak{g} \subset \mathfrak{so}(N, \mathbb{C})$ be an irreducible representation of real type of a complex simple Lie algebra different from $\mathfrak{sl}(2, \mathbb{C})$, with $0 \notin \Omega$ and satisfying (SII). Then the roots system and the highest weight of the representation is is one of the following (modulo congruence):*

A_n: ω_4 for $n = 7$.

B_n: ω_n for $n = 3, 4, 7$.

D_n: ω_1, $2\omega_1$ for arbitrary n and ω_8 for $n = 8$.

Proof. We apply proposition 4.42 and corollary 4.46 to the remaining representations with $0 \notin \Omega$, i.e. representations of A_n, B_n, C_n, D_n, E_6 and E_7. Therefore we use a fundamental system such that the extremal weight Λ determined by (SII) is the highest weight. It can be written in the fundamental representations $\Lambda = \sum_{k=1}^{n} m_k \omega_k$ with $m_k \in \mathbb{N} \cup \{0\}$.

A_n: Proposition 4.42 gives for the largest root

$$2 \geq \frac{2\langle \Lambda, e_1 - e_{n+1} \rangle}{\|e_1 - e_{n+1}\|^2} = \sum_{k=1}^{n} m_k \langle \omega_k, e_1 - e_{n+1} \rangle = \sum_{k=1}^{n} m_k.$$

Since the representation has to be self dual we have that $m_i = m_{n+1-i}$.

First we consider the case that $\Lambda = \omega_i + \omega_{n+1-i}$. For $n > 2$ we get in case $i > 1$ that $\langle \Lambda, e_2 - e_n \rangle = 2$. But $(e_2 - e_n) \perp (e_1 - e_{n+1})$ gives a contradiction to 1 of corollary 4.46. For $n \geq 2$ it has to be

$$\Lambda = \omega_1 + \omega_n = 2e_1 + e_2 + \ldots e_n = e_1 - e_{n+1}$$

recalling that for A_n holds that $e_1 = -(e_2 + \ldots + e_{n+1})$. Thus the representation is the adjoint one with $0 \in \Omega$.

Now we consider the case that $n + 1$ is even and $\Lambda = 2\omega_{\frac{n+1}{2}}$. This representation is orthogonal but again we have $\langle \Lambda, e_2 - e_n \rangle = 2$ for $n > 2$. But this is impossible because of point 1 of corollary 4.46.

For $n + 1$ even we have to study the case $\Lambda = \omega_{\frac{n+1}{2}}$. This representation is orthogonal if $\frac{n+1}{2}$ is even. The weights of this representation are given by $\pm e_{k_1} \pm \ldots \pm e_{k_{\frac{n+1}{2}}}$ where the $\pm$'s are meant to be independent of each other.

We will show that (SII) implies $n \leq 7$.

Hence suppose that there is a root α such that (SII) with Λ. We have to consider two cases for α. The first is that $\alpha = e_i - e_j$ with $1 \leq i \leq \frac{n+1}{2} < j \leq n + 1$. W.l.o.g. we take $\alpha = e_{\frac{n+1}{2}} - e_{\frac{n+1}{2}+1}$ and consider the weight

$$\lambda := e_1 + \ldots e_{\frac{n+1}{2}-3} + e_{\frac{n+1}{2}+1} + e_{\frac{n+1}{2}+2} + e_{\frac{n+1}{2}+3}.$$

$\langle \lambda, \alpha \rangle < 0$ implies $\lambda \in \Omega_\alpha$. Then $\lambda - (\Lambda - \alpha) \in \Delta_0$ or $\lambda + \Lambda \in \Delta_0$. We check the first alternative: $\Lambda - \alpha = e_1 + \ldots e_{\frac{n+1}{2}-1} + e_{\frac{n+1}{2}+1}$ implies

$$\lambda - (\Lambda - \alpha) = e_{\frac{n+1}{2}-3} + e_{\frac{n+1}{2}-2} + e_{\frac{n+1}{2}+2} + e_{\frac{n+1}{2}+3}.$$

But this is not a root.

For the second alternative we get, recalling that $-e_1 = e_2 + \ldots + e_{n+1}$,

$$\lambda + \Lambda = e_1 + \ldots + e_{\frac{n+1}{2}-3} - e_{\frac{n+1}{2}+4} - \ldots - e_{n+1}.$$

This is not a root if $\frac{n+1}{2} > 4$, i.e. $n > 7$.

For the second type of root $\alpha = e_i - e_j$ with $1 \leq i < j \leq \frac{n+1}{2}$ and $\frac{n+1}{2} < i < j \leq n+1$ one derives analogously that $n \leq 5$.

Hence for $\Lambda = \omega_{\frac{n+1}{2}}$ the property (SII) can only be fulfilled if $n \leq 7$. These representations are orthogonal for $n = 7$ and $n = 3$. A_3 is isomorphic to D_3 and the representation with highest weight ω_2 of A_3 is equivalent to the one with ω_1 of D_3.

B_n: Again proposition 4.42 gives for the largest root

$$2 \geq \frac{2\langle \Lambda, e_1 + e_2 \rangle}{\|e_1 + e_2\|^2} = \sum_{k=1}^{n} m_k \langle \omega_k, e_1 + e_2 \rangle = m_1 + 2m_2 + \ldots 2m_{n-1} + m_n.$$

The only representations with $0 \notin \Omega$ are these with $\Lambda = \omega_1 + \omega_n$ and the spin representation $\Lambda = \omega_n$. There is no possibility to apply the first point of corollary 4.46. But we verify that for $\Lambda = \omega_1 + \omega_n$ (SII) implies $n \leq 2$ and for the spin representation $\Lambda = \omega_n$ (SII) implies $n \leq 7$.

The spin representations: For these we show that (SII) implies $n \leq 7$ The spin representation of highest weight $\Lambda = \frac{1}{2}(e_1 + \ldots + e_n)$ has weights $\Omega = \left\{\frac{1}{2}(\varepsilon_1 e_1 + \ldots + \varepsilon_n e_n) | \varepsilon_i = \pm 1\right\}$. We have to consider three types for the root α: $\alpha = e_i$, $\alpha = e_i + e_j$ and $\alpha = e_i - e_j$.

For the first we can assume w.l.o.g. that $\alpha = e_1$. Then $\Omega_\alpha = \{\frac{1}{2}(-e_1 + \varepsilon_2 e_2 + \ldots + \varepsilon_n e_n) | \varepsilon = \pm 1\}$. It is $\Lambda - \alpha = \frac{1}{2}(-e_1 + e_2 + \ldots + e_n)$. Hence for $\lambda \in \Omega_\alpha$ we have

$$\begin{aligned} \Lambda - \alpha - \lambda &= \frac{1}{2}((1 - \varepsilon_2)e_2 + \ldots + (1 - \varepsilon_n)e_n) \text{ and} \\ \Lambda + \lambda &= \frac{1}{2}((1 + \varepsilon_2)e_2 + \ldots + (1 + \varepsilon_n)e_n) \end{aligned}$$

If (SII) is satisfied at least one of these expression has to be a root. But if $n \geq 7$ we can choose $(\varepsilon_2, \ldots \varepsilon_n)$ such that non of them is a root.

The second type of root shall be w.l.o.g. $\alpha = e_1 - e_2$. In this case $\Omega_\alpha = \{\frac{1}{2}(-e_1 + e_2 + \varepsilon_3 e_3 + \ldots + \varepsilon_n e_n) | \varepsilon_i = \pm 1\}$ and $\Lambda - \alpha = \frac{1}{2}(-e_1 + 2e_2 + e_3 + \ldots + e_n)$. Hence for $\lambda \in \Omega_\alpha$

$$\begin{aligned} \Lambda - \alpha - \lambda &= \frac{1}{2}(e_2 + (1-\varepsilon_3)e_3 + \ldots + (1-\varepsilon_n)e_n) \text{ and} \\ \Lambda + \lambda &= \frac{1}{2}(2e_2 + (1+\varepsilon_3)e_3 + \ldots + (1+\varepsilon_n)e_n) \end{aligned}$$

We can choose λ such that none of them is a roots if $n \geq 4$.

Now we consider the last type of root, $\alpha = e_1 + e_2$. $\Omega_\alpha = \{\frac{1}{2}(-e_1 - e_2 + \varepsilon_3 e_3 + \varepsilon_n e_n) | \varepsilon_i = \pm 1\}$ and $\Lambda - \alpha = \frac{1}{2}(-e_1 - e_2 + e_3 + \ldots + \ldots + e_n)$. Hence for $\lambda \in \Omega_\alpha$

$$\begin{aligned} \Lambda - \alpha - \lambda &= \frac{1}{2}((1-\varepsilon_3)e_3 + \ldots + (1-\varepsilon_n)e_n) \text{ and} \\ \Lambda + \lambda &= \frac{1}{2}((1+\varepsilon_3)e_3 + \ldots + (1+\varepsilon_n)e_n) \end{aligned}$$

We can choose λ such that none of them is a roots if $n \geq 8$. Hence if (SII) is satisfied it has to be $n \leq 7$ and for $n = 7$ the pair of property (SII) is of the shape $(\Lambda, e_1 + e_2)$.

Now for $n = 2$, $n = 5$ and $n = 6$ the spin representations are symplectic but not orthogonal.

The representations of $\Lambda = \omega_1 + \omega_n = \frac{3}{2}e_1 + \frac{1}{2}(e_2 + \ldots + e_n)$. Then the weights are given by $\frac{1}{2}(a e_{k_1} + \varepsilon_2 e_{k_2} + \ldots + \varepsilon_n e_{k_n})$ with $a \in \{\pm 1, \pm 3\}$ and $\varepsilon_i = \pm 1$. For these one shows analogously that (SII) implies $n \leq 2$. For $n = 2$ this representation is symplectic.

C_n: For the largest root we get

$$2 \geq \frac{2\langle \Lambda, 2e_1 \rangle}{\|2e_1\|^2} = \sum_{k=1}^{n} m_k \langle \omega_k, e_1 \rangle = \sum_{k=1}^{n} m_k.$$

In case that one $m_i = 2$ and all others zero we have that $0 \in \Omega$. Hence we suppose that $\Lambda = \omega_i + \omega_j$ for $i \neq j$. If $i > 1$ we get for the root $2e_2$ which is orthogonal to $2e_1$ that $\frac{2\langle \Lambda, 2e_2 \rangle}{\|2e_2\|^2} = 2$. Thus by 1 of corollary 4.46 we have $i = 1$. But $\Lambda = \omega_1 + \omega_i$ is only orthogonal if i is odd, but if i is odd we have that $0 \in \Omega$.

Hence we have to deal with the case $\Lambda = \omega_i$. This is orthogonal if i is even, but in this case $0 \in \Omega$.

D_n: Here we get for the largest root

$$2 \geq \frac{2\langle \Lambda, e_1 + e_2 \rangle}{\|e_1 + e_2\|^2} = \sum_{k=1}^{n} m_k \langle \omega_k, e_1 + e_2 \rangle = m_1 + 2m_2 + \ldots + 2m_{n-2} + m_{n-1} + m_n.$$

First we consider the representation where this number is equal to 2.

For the representations $2\omega_n$ and $2\omega_{n-1}$ we have that $0 \in \Omega$.

For the representations $\Lambda = \omega_1 + \omega_n$ and $\Lambda = \omega_1 + \omega_{n-1}$ we get that $n = 4$ or there is no triple as in the second point of proposition 4.46. Thus suppose in this case $n > 4$. We have that$\langle \Lambda, e_1 + e_2 \rangle = 2$ and for the orthogonal roots $\langle \Lambda, e_1 - e_2 \rangle = \langle \Lambda, e_3 \pm e_4 \rangle = 1$. But this contradicts proposition 4.46,1.

For $\Lambda = \omega_{n-1} + \omega_n = e_1 + \ldots + e_{n-1}$ we have that $0 \notin \Omega$ implies $n-1$ even. The first point of corollary 4.46 then gives for $n > 4$ that $2 = \langle \Lambda, e_3 + e_4 \rangle$ which is impossible. Hence $n \leq 4$. Then $1 = \langle \Lambda, e_3 \pm e_4 \rangle$ and the second point of corollary 4.46 imply $n \leq 3$.

Now suppose that $\Lambda = \omega_i$ for $2 \leq i \leq n-2$. We apply the first point of corollary 4.46. If $n \geq 4$ we get that $\langle \omega_i, e_3 + e_4 \rangle = 2$ for $i \geq 4$ but this was excluded. Hence $i \leq 3$.

In the case $n = 3$ we have that only ω_2 is an orthogonal representation. But for this holds that $0 \in \Omega$.

Thus, to get the assertion of the proposition we have to show that

1. For the spin representations $\Lambda = \omega_{n-1}$ and $\Lambda = \omega_n$ (SII) implies $n \leq 8$
2. $\Lambda = \omega_3$ does not satisfy (SII),
3. $\Lambda = \omega_1 + \omega_3$ and $\omega_1 + \omega_4$ for $n = 4$ do not satisfy (SII).

The spin representations: For these we show that (SII) implies first $n \leq 8$. Because we are interested in the representations modulo congruence it suffices to consider the spin representation of highest weight $\Lambda = \frac{1}{2}(e_1 + \ldots + e_n)$ with weights $\Omega = \left\{\frac{1}{2}(\varepsilon_1 e_1 + \varepsilon_n e_n) | \varepsilon_i = \pm 1 \text{ and } \varepsilon_i = -1 \text{ for an even number}\right\}$.

Analogously as for B_n we get for two types of roots $\alpha = e_i + e_j$ and $\alpha = e_i - e_j$ that (SII) implies $n \leq 8$ (We have to admit one dimension higher because of the sign restriction of the weights).

Now for n odd the spin representation is not self dual, and for $n = 6$ not orthogonal. For $n = 4$ it is congruent to ω_1.

$\Lambda = \omega_3 = e_1 + e_2 + e_3$: Here it is $\Omega = \{(\varepsilon_1 e_{k_1} + \varepsilon_2 e_{k_2} + \varepsilon_3 e_{k_3} | \varepsilon_i = \pm 1\} \cup \{\pm e_i\}$. For $n = 3$ and $n = 4$ this is a spin representation. Hence suppose $n \geq 5$.

For $\alpha = e_1 + e_2$ we get $\Lambda - \alpha = e_3$. Set $\lambda := -e_1 + e_4 + e_5 \in \Omega_\alpha$. Hence $\Lambda - \alpha - \lambda = e_3 + e_1 - e_4 - e_5$ and $\Lambda + \lambda = e_2 + e_3 + e_4 + e_5$. None is a root, i.e. ω_3 for $n \geq 5$ does not satisfy (SII).

For $\alpha = e_1 - e_2$ we get the same.

$\Lambda = \omega_1 + \omega_3$ *and* $\omega_1 + \omega_4$ *for* $n = 4$. These are congruent to each other and as above it can be shown that they do not satisfy (SII).

E_6 and E_7: For these we refer to [Sch99]. There is shown that under the conclusions of proposition 4.42 and 4.43 — which is our situation because of lemma 4.47 — the only remaining representations are the standard representations of E_6 and E_7. But the first is not self dual and the latter symplectic but not orthogonal.

□

We get the following

4.49 Corollary. *Let* $\mathfrak{g} \subset \mathfrak{so}(N, \mathbb{C})$ *be an orthogonal algebra of real type different from* $\mathfrak{sl}(2, \mathbb{C})$. *If* $0 \notin \Omega$ *and (SII) is satisfied, then it is the complexification of a Riemannian holonomy representation or the spin representation of* $\mathfrak{so}(15, \mathbb{C})$.

Proof. We give the Riemannian manifolds the complexified holonomy representation of which is one of the representations of proposition 4.48.
The representation with highest weight ω_4 of A_7 is the complexified holonomy representation of the symmetric space of type EV, i.e. of $E_7/SU(8)$ resp. $E_{7(7)}/SU(8)$.
The spin representations of B_n for $n = 3, 4$ are the holonomy representations of a non-symmetric $Spin(7)$–manifold and of the symmetric space of type FII, i.e. of $F_4/Spin(9)$ resp. $F_{4(-20)}/Spin(9)$. For $n = 7$ we have an exception.
For D_n first we have the standard representation, i.e. the complexified holonomy representation of a generic manifold. The representation with highest weight $2\omega_1$ is the complexified holonomy representation of the symmetric space of type AI, i.e. of $SU(2n)/SO(2n, \mathbb{R})$, resp. $Sl(2n, \mathbb{R})/SO(2n, \mathbb{R})$. The remaining representation of $Spin(16)$ is the complexified holonomy representation of the symmetric space of type $EVIII$, i.e. of $E_8/Spin(16)$, resp. $E_{8(8)}/Spin(16)$. □

4.3.5 Consequences for simple weak-Berger algebras of real type

Before we conclude the result we need a lemma to exclude both exceptions.

4.50 Lemma. *The spin representation of* B_7 *and the representation of* G_2 *with two times a short root as highest weight are not weak-Berger.*

Proof. 1.) Suppose that the spin representation of B_7 is weak-Berger. We have shown that it does not satisfy the property (SI). Hence it obeys (SII). Let (Λ, α) be the pair of (SII). We choose a fundamental system such that $\Lambda = \omega_7$ is the highest weight. In the proof of proposition 4.48 we have shown that in this case $\alpha = e_i + e_j$.
Let now Q_ϕ be the weight element from $\mathcal{B}_H(\mathfrak{g})$ and $u_\Lambda \in V_\Lambda$ such that $Q_\phi(u_\Lambda) = A_{e_i+e_j} \in \mathfrak{g}_{e_i+e_j}$. Since $Q_\phi(u_\Lambda) \in \mathfrak{g}_{\phi+\Lambda}$ this implies that $\phi = e_i + e_j - \Lambda$ is a weight of

$\mathcal{B}_H(\mathfrak{g})$. Hence $\phi = -\frac{1}{2}(e_1 + \ldots + e_{i-1} - e_i + e_{i+1} + \ldots + e_{j-1} - e_j + e_{j+1} + \ldots + e_7)$ is also an extremal weight of V and we can consider a weight vector $u_{-\phi} \in V_{-\phi}$. For this we get $Q_\phi(u_{-\phi}) \in \mathfrak{t}$. In case it does not vanish it would define a planar spanning triple $(\phi, -\phi, (Q_\phi(u_{-\phi}))^\perp)$, i.e. (SI) would be satisfied. But this was not possible, and thus $Q_\phi(u_{-\phi}) = 0$.
On the other hand we have that $0 \neq Q_\phi(u_\Lambda)u_{-\phi} \in V_\Lambda$ and thus there is a $v \in V_{-\Lambda}$ such that $H(Q_\phi(u_\Lambda)u_{-\phi}, v) \neq 0$. Now the Bianchi identity gives

$$0 = H(Q_\phi(u_\Lambda)u_{-\phi}, v) + \underbrace{H(Q_\phi(u_{-\phi})v, u_\Lambda))}_{=0} + H(Q_\phi(v)u_\Lambda, u_{-\phi}, v).$$

Hence $0 \neq Q_\phi(v) \in \mathfrak{g}_{\phi-\Lambda}$. But $\phi - \Lambda = -(e_1 + \ldots + e_{i-1} + e_{i+1} + \ldots + e_{j-1} + e_{j+1} + \ldots + e_7)$ is not a root, hence $\mathfrak{g}_{\phi-\Lambda} = \{0\}$. This is a contradiction.
2.) Suppose that the representation of G_2 with two times a short root as highest weight is weak-Berger. We will argue analogously as for B_n.

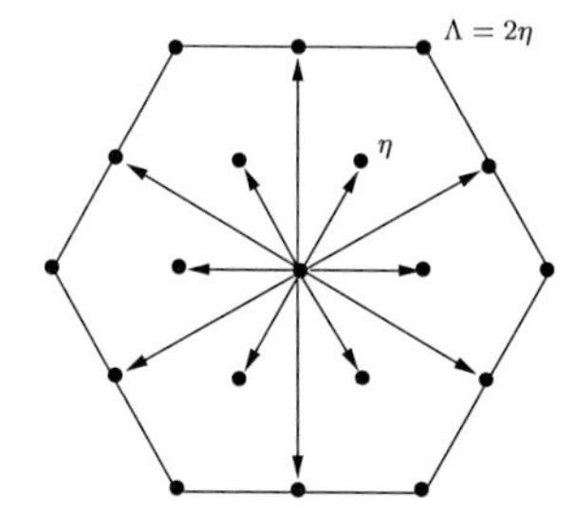

In the picture we see the weight lattice of this representation. Obviously there is no planar spanning triple, because there is no hypersurface which contains all but two extremal weight (see also proof of proposition 4.31).
The weak-Berger property implies that there is a pair (Λ, α) such that (SII) is satisfied. We choose a fundamental system such that $\Lambda = 2\eta$ is the maximal weight.
Using the realization of G_2 from the appendix of [Kna02] we have that $\eta = e_3 - e_2$. Now we have to determine the roots for which (SII) is satisfied.

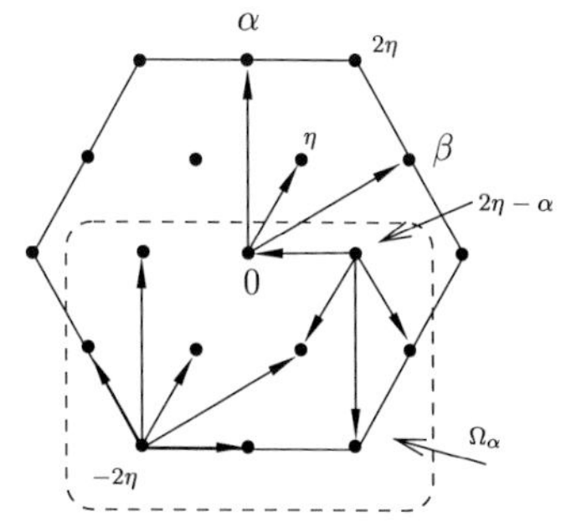

In the picture one can see that the long roots α and β satisfy (SII). (We illustrate the situation in detail only for α.) Contemplate the picture for a moment one sees that there are no short roots and no other long root for which (SII) can be valid.
Now α and β are the only roots with $\langle \Lambda, \alpha \rangle > 0$ and $\langle \Lambda, \beta \rangle > 0$. Hence $\alpha = 2e_3 - e_1 - e_2$ and $\beta = -2e_2 + e_1 + e_3$.

We consider the case where (Λ, α) satisfies (SII). There is a weight element Q_ϕ from $\mathcal{B}_H(\mathfrak{g})$ such that $Q_\phi(u_\Lambda) = A_{2e_3 - e_1 - e_2}$, i.e. $\phi = 2e_3 - e_1 - e_2 - \Lambda = e_2 - e_1$. But this is a

short root and therefore a weight. Thus we consider $u_{-\phi} \in= V_{-\phi}$. Then $Q_\phi(u_{-\phi}) \in \mathfrak{t}$. Since there is no planar spanning triple it has to be zero. As above the Bianchi identity gives that $\phi - \Lambda$ has to be a root. But $\phi - \Lambda = e_2 - e_1 - 2e_3 + 2e_2 = 3e_2 - 2e_3 - e_1$ is no root.
For β one proceeds analogously. □

Now we can draw the conclusions from the previous sections. If a Lie algebra acts irreducible of real type the it is semi-simple and obeys the properties (SI) or (SII). The simple Lie algebras with (SI) or (SII) we have listed above. Thus we get

4.51 Theorem. *Let $\mathfrak{g} \subset \mathfrak{so}(N,\mathbb{R})$ be a irreducible simple weak-Berger algebra of real type. Then it is the holonomy representation of a Riemannian manifold.*
The conclusion holds in particular if $\mathfrak{g}$ is simple, of real type and the irreducible component of the $\mathfrak{so}(n)$-projection of an indecomposable, non-irreducible simply connected Lorentzian manifold.

4.52 Remark. Quaternionic symmetric spaces. With the result of course we have covered all simple irreducible acting Riemannian holonomy groups of real type.
If one considers a quaternionic symmetric space $G/Sp(1) \cdot H$ with $H \subset Sp(n)$ then of course $\mathfrak{sp}(1) \oplus \mathfrak{h} \subset \mathfrak{so}(4n,\mathbb{R})$ is a real Berger algebra of real type and thus its complexification is a complex Berger algebra of real type. Then the restriction of this representation to $\mathfrak{h}$ is of quaternionic, i.e. of non-real type, its complexification decomposes into two irreducible components $\mathbb{C}^{2n} \oplus \overline{\mathbb{C}^{2n}}$. For this situation in [Sch99] is proved that $\mathfrak{h}^{\mathbb{C}}_{|\mathbb{C}^{2n}}$ is a complex Berger algebra. This result does not collide with our list because this representation is not of real type and hence not orthogonal. $\mathfrak{h} \subset \mathfrak{so}(4n,\mathbb{R})$ is not a real Berger algebra.

4.4 Weak-Berger algebras of non-real type

In this section we will classify the weak-Berger algebras of non-real type, and we will show that these are Berger algebras. For the classification we will use the classification of first prolongations of irreducible complex Lie algebras. We will show that the complexification of the space $\mathcal{B}_h(\mathfrak{g}_0)$ is isomorphic to the first prolongation of the complexified Lie algebra.

In this section $\mathfrak{g}_0$ is a real Lie algebra and E a $\mathfrak{g}_0$-module of non-real type, i.e. $E^{\mathbb{C}}$ is not irreducible. Thus the situation is a little bit more puzzling then in the real case. Since $\mathfrak{g}_0 \subset \mathfrak{so}(E,h)$ with h positive definite, $\mathfrak{g}_0$ is compact. For a compact real Lie algebra with module of non-real type the corresponding complex representation of non-real type is not orthogonal but unitary (See appendix B, in particular proposition B.21).

But if we switch to the complexified algebra the $(\mathfrak{g}^{\mathbb{C}}, V)$ irreducible remains, but it can no longer be unitary of course. We have to handle this situation.
With the same notations as in appendix B the complex representations space $W = E^{\mathbb{C}}$ splits into the irreducible modules $W = V \oplus \overline{V}$ under $\mathfrak{g}_0$. This splitting is of course $\mathfrak{g}_0^{\mathbb{C}}$ invariant.
Now we define the complex Lie algebra

$$\mathfrak{g} \;:=\; \left\{ A_{|V} \;\middle|\; A \in \mathfrak{g}_0^{\mathbb{C}} \subset \mathfrak{so}(W = V \oplus \overline{V}, H) \right\} \subset \mathfrak{gl}(V). \tag{4.56}$$

Here H denotes again $h^{\mathbb{C}}$. Since the symmetric bilinear form we start with is positive definite the appendix B gives two important results (see proposition B.21):

1. Since $\mathfrak{g}_0$ is compact there is a positive definite hermitian form θ^h on V which is the the restriction of the sesqui-linear extension of h on V, such that $(\mathfrak{g}_0)_{|V} \subset \mathfrak{u}(V, \theta^h)$.

2. $\mathfrak{g}$ is not orthogonal, in particular $H_{|V \times V} = 0$. This is the case since modules of non-real type are symplectic if they are self-dual. Thus they cannot be orthogonal.

In $\mathfrak{g}_0^{\mathbb{C}}$ as well as in $\mathfrak{g}$ we have a conjugation $\overline{}$ with respect to $\mathfrak{g}_0$ and $(\mathfrak{g}_0)_{|V}$ respectively. Since an $A \in \mathfrak{g}_0$ acts on $V \oplus \overline{V}$ by $A(v + \overline{w}) = Av + \overline{Aw}$ we have for $iA \in \mathfrak{g}_0^{\mathbb{C}}$ that

$$iA(v + \overline{w}) \;=\; i(Av + \overline{Aw}) \;=\; (iAv + \overline{-iAw}).$$

So we write the action of $A \in \mathfrak{g}_0^{\mathbb{C}}$ with the help of the conjugation in $\mathfrak{g}$ as follows

$$A(v + \overline{w}) \;=\; Av + \overline{\overline{A}w}. \tag{4.57}$$

This gives the following Lie algebra isomorphism

$$\begin{array}{rccc} \varphi \;: & \mathfrak{g}_0^{\mathbb{C}} & \simeq & \mathfrak{g} \\ & A & \mapsto & A_{|V}. \end{array}$$

This is clearly a Lie algebra homomorphism. It is injective because for $A_{|V} = B_{|V}$ holds that $A(v + \overline{w}) = Av + \overline{\overline{A}w} = Bv + \overline{\overline{B}w} = B(v + \overline{w})$ for all $v, w \in V$, i.e. $A = B$.
By definition it is surjective and φ^{-1} is given by

$$\varphi^{-1}(A) \;:\; v + \overline{w} \;\longmapsto\; Av + \overline{\overline{A}w} \quad \text{for all} \quad A \in \mathfrak{g}. \tag{4.58}$$

These notations are needed to show the relation to the first prolongation.

4.4.1 The first prolongation of a Lie algebra of non-real type

Now we define the first prolongation of an arbitrary Lie algebra $\mathfrak{g} \subset \mathfrak{gl}(V)$.

4.53 Definition. The $\mathfrak{g}$-module

$$\mathfrak{g}^{(1)} := \{Q \in V^* \otimes \mathfrak{g} \mid Q(u)v = Q(v)u\}. \tag{4.59}$$

is called **first prolongation** of $\mathfrak{g} \subset \mathfrak{gl}(V)$. Furthermore we set

$$\tilde{\mathfrak{g}} := span\{Q(u) \in \mathfrak{g} \mid Q \in \mathfrak{g}^{(1)}, u \in V\} \subset \mathfrak{g},$$

and if in $\mathfrak{g}$ a conjugation $^-$ is given:

$$\begin{aligned} \mathfrak{g}^{[1,1]} &:= \{R \in \overline{V}^* \otimes \mathfrak{g}^{(1)} \mid \overline{R(\overline{u}, v)} = -R(\overline{v}, u)\}, \\ \tilde{\tilde{\mathfrak{g}}} &:= span\{R(\overline{u}, v) \mid R \in \mathfrak{g}^{[1,1]}, \overline{u} \in \overline{V}, v \in V\} \subset \mathfrak{g}. \end{aligned}$$

We will now describe the spaces $\mathcal{B}_H(\mathfrak{g}_0^{\mathbb{C}})$ and $\mathcal{K}(\mathfrak{g}_0^{\mathbb{C}})$ — which are essential for the Berger and the weak-Berger property — with the help of the first prolongation of $\mathfrak{g}$.
In the setting of the above notations we can now prove the following.

4.54 Proposition. *Let E be a non-real type module of $\mathfrak{g}_0$, orthogonal with respect to a positive definite scalar product h, and $E^{\mathbb{C}} = V \oplus \overline{V}$ the corresponding $\mathfrak{g}_0^{\mathbb{C}}$ invariant decomposition, $\mathfrak{g}$ defined as in (4.56). Then there is an isomorphism*

$$\begin{aligned} \phi \; : \; \mathcal{B}_H(\mathfrak{g}_0^{\mathbb{C}}) &\simeq \mathfrak{g}^{(1)} \\ Q &\mapsto Q_{|V \times V}. \end{aligned}$$

Proof. For the prove we will use the $\mathfrak{g}_0$–invariant hermitian form θ on V which is given by $\theta(u,v) = h^{\mathbb{C}}(u, \overline{v})$, where $^-$ is the conjugation in $E^{\mathbb{C}} = V \oplus \overline{V}$ with respect to E. The linearity of ϕ mapping is clear. we have to show the following:
1.) The definition of ϕ is correct, i.e. for $Q \in \mathcal{B}_H(\mathfrak{g}_0^{\mathbb{C}})$ it is $Q_{|V\times V} \in \mathfrak{g}^{(1)}$. We have for every $u, v, w \in V$ and $H = h^{\mathbb{C}}$ that

$$\begin{aligned} \theta(Q(u)v, w) &= h^{\mathbb{C}}(Q(u)v, \overline{w}) \\ &= -h^{\mathbb{C}}(Q(v)\overline{w}, u) - \underbrace{h^{\mathbb{C}}(Q(\overline{w})u, v)}_{\substack{=0 \\ \text{since } h^{\mathbb{C}}_{V\times V} = 0 \text{ (proposition B.21)}}} \\ &\overset{h^{\mathbb{C}} \text{ invariant}}{=} h^{\mathbb{C}}(Q(v)u, \overline{w}) \\ &= \theta(Q(v)u, w), \end{aligned}$$

i.e. $Q(u)v = Q(v)u$ which means that $Q_{|V\times V} \in \mathfrak{g}^{(1)}$.
2.) The homomorphism ϕ is injective. Let Q_1 and Q_2 be in $\mathcal{B}_H(\mathfrak{g}_0^{\mathbb{C}})$ with $(Q_1)_{|V\times V} = (Q_2)_{|V\times V}$. Then it is

a) $(Q_1)_{|\overline{V}\times\overline{V}} = (Q_2)_{|\overline{V}\times\overline{V}}$, since $Q_1(\overline{u})\overline{v} = \overline{Q_1(u)v} = \overline{Q_2(u)v} = Q_2(\overline{u})\overline{v}$,

b) $(Q_1)_{|\overline{V}\times V} = (Q_2)_{|\overline{V}\times V}$, since

$$\begin{aligned} \theta(Q_1(\overline{u})v, w) &= h^{\mathbb{C}}(Q_1(\overline{u})v, \overline{w}) = -h^{\mathbb{C}}(v, Q_1(\overline{u})\overline{w}) = \\ = h^{\mathbb{C}}(v, Q_2(\overline{u})\overline{w}) &= h^{\mathbb{C}}(Q_2(\overline{u})v, \overline{w}) = \theta(Q_2(\overline{u})v, w). \end{aligned}$$

c) $(Q_1)_{|V\times\overline{V}} = (Q_2)_{|V\times\overline{V}}$ because of b) with the same argument as in a).

3.) The homomorphism ϕ is surjective. For $Q \in \mathfrak{g}^{(1)}$ we define ϕ^{-1} using φ:

$$\begin{aligned} &(\phi^{-1}Q)(u) := \varphi^{-1}(Q(u)) \quad \text{and} \quad (\phi^{-1}Q)(\overline{u}) := \varphi^{-1}(\overline{Q(u)}) \in \mathfrak{gl}(E^{\mathbb{C}}), \\ &\text{i.e. } (\phi^{-1}Q)(u,v) = Q(u)v \quad , \quad (\phi^{-1}Q)(\overline{u}) = \overline{Q(u)}v \ , \\ &(\phi^{-1}Q)(u,\overline{v}) = \overline{\overline{Q(u)}v} \quad , \quad (\phi^{-1}Q)(\overline{u},\overline{v}) = \overline{Q(u)v}. \end{aligned}$$

It is $(\phi^{-1}Q)(\overline{u},\overline{v}) = \overline{(\phi^{-1}Q)(u,v)}$.
Then obviously $\phi \circ \phi^{-1} = id$, since $\phi\left(\phi^{-1}(Q)\right) = \phi^{-1}(Q)_{|V\times V} = Q$.
Because of the symmetry of Q we have also that $(\phi^{-1}Q) \in \mathcal{B}_H(\mathfrak{g}_0^{\mathbb{C}})$:

- For $u, v \in V, \overline{w} \in \overline{V}$:

$$\begin{aligned} H((\phi^{-1}Q)(u)v,\overline{w}) + H((\phi^{-1}Q)(v)\overline{w},u) + \overbrace{H((\phi^{-1}Q)(\overline{w})u,v)}^{=0 \text{ because } H=0 \text{ on } V\times V} &= \\ H((\varphi^{-1}(Q(u))v,\overline{w}) + \underbrace{H((\varphi^{-1}(Q(v))\overline{w},u)}_{=-H((\overline{w},\varphi^{-1}(Q(v))u)} &= H((Q(u)v - Q(v)u,\overline{w}) \\ &= 0. \end{aligned}$$

- For $u \in V, \overline{v}, \overline{w} \in \overline{V}$:

$$\begin{aligned} \overbrace{H((\phi^{-1}Q)(u)\overline{v},\overline{w})}^{=0} + H((\phi^{-1}Q)(\overline{v})\overline{w},u) + H((\phi^{-1}Q)(\overline{w})u,\overline{v}) &= \\ H((\varphi^{-1}(\overline{Q(v)})\overline{w},u) + \underbrace{H(\varphi^{-1}(\overline{Q(w)})u,\overline{w})}_{=-H((u,\varphi^{-1}(\overline{Q(w)})\overline{v})} &= H(\overline{(Q(v)w - Q(w)v}, u) \\ &= 0. \end{aligned}$$

Terms with entries only from V or only from $\overline{V}$ are zero. □

Furthermore we show for the space $\mathcal{K}(\mathfrak{g})$ an analogous result.

4.55 Proposition. *Let E be an orthogonal non-real type module of $\mathfrak{g}_0$ and $E^{\mathbb{C}} = V \oplus \overline{V}$ the corresponding $\mathfrak{g}_0^{\mathbb{C}}$ invariant decomposition, $\mathfrak{g}$ defined as in (4.56). Suppose that $\theta := \theta^h$ is non-degenerate. Then there is an isomorphism*

$$\begin{aligned} \psi \ : \ \mathcal{K}(\mathfrak{g}_0^{\mathbb{C}}) &\simeq \mathfrak{g}^{[1,1]} \\ R &\mapsto R_{|\overline{V}\times V\times V}. \end{aligned}$$

Proof. The proof is completely analogous to the previous one.
1.) The definition is correct. We have for $u, v, w \in V$ and $R \in \mathcal{K}(\mathfrak{g}_0^{\mathbb{C}})$ that

$$\underbrace{R(u,v)\overline{w}}_{\in \overline{V}} = \underbrace{R(\overline{w},v)u}_{\in V} - \underbrace{R(\overline{w},u)v}_{\in V} = 0.$$

but this means that $R(\overline{u},.)_{|V\times V} \in \mathfrak{g}^{(1)}$.
Further $R(u,v)\overline{w} = 0$ implies $R(u,v)w = 0$ because

$$\theta(R(u,v)w,z) = h^{\mathbb{C}}(R(u,v)w,\overline{z}) = -h^{\mathbb{C}}(w,R(u,v)\overline{z}) = 0.$$

This implies $R(\overline{u},\overline{v})\overline{w} = R(\overline{u},\overline{v})w = 0$ too.
For a $R \in \mathcal{K}(\mathfrak{g}_0^{\mathbb{C}})$ we have due to the skew symmetry

$$\overline{R(\overline{u},v)} \overset{\text{easy calculation}}{=} R(u,\overline{v}) \overset{\text{skew-symm.}}{=} -R(\overline{v},u),$$

i.e. the restriction of R on $\overline{V} \times V \times V$ is in $\mathfrak{g}^{[1,1]}$.
2.) The homomorphism ψ is injective.
Let R_1 and R_2 be in $\mathcal{K}(\mathfrak{g}_0^{\mathbb{C}})$ with $(R_1)_{\overline{V}\times V\times V} = (R_2)_{\overline{V}\times V\times V}$. Then again via θ the remaining non zero terms $R_i(\overline{u},v)\overline{w}$ are determined by $R_i(\overline{u},v)w$ which are equal for $i = 1,2$ and by the skew symmetry of R.
3.) The homomorphism ψ is surjective.
We set

$$\begin{aligned}(\psi^{-1}R)(\overline{u},v)) &:= \varphi^{-1}(R(\overline{u},v) \quad , \quad (\psi^{-1}R)(u,\overline{v}) := \varphi^{-1}(\overline{R(\overline{u},v)}) \text{ and} \\ (\psi^{-1}R)(u,v) &:= (\psi^{-1}R)(\overline{u},\overline{v}) := 0\end{aligned}$$

So we have the skew symmetry, i.e. $\psi^{-1}R \in \wedge^2 E^{\mathbb{C}} \otimes \mathfrak{g}_0^{\mathbb{C}}$, because

$$(\psi^{-1}R)(u,\overline{v}) = \varphi^{-1}(\overline{R(\overline{u},v)}) = -\varphi^{-1}(R(\overline{v},u)) = -(\psi^{-1}R)(\overline{v},u).$$

The Bianchi identity is also satisfied:

- For $u \in \overline{V}, v, w \in V$:

$$\begin{aligned}(\psi^{-1}R)(\overline{u},v)w + \overbrace{(\psi^{-1}R)(v,w)\overline{u}}^{=0} + (\psi^{-1}R)(w,\overline{u})v &= \\ \varphi^{-1}\left(R(\overline{u},v)\right)w + \varphi^{-1}\left(\overline{R(\overline{w},u)}\right)v &= \\ \varphi^{-1}\left(R(\overline{u},v)\right)w - \varphi^{-1}\left(R(\overline{u},w)\right)v &= \\ R(\overline{u},v)w - R(\overline{u},w)v &= 0\end{aligned}$$

• For $\overline{u}, \overline{v} \in \overline{V}, w \in V$:

$$\begin{aligned}\overbrace{(\psi^{-1}R)(\overline{u},\overline{v})w}^{=0}+(\psi^{-1}R)(\overline{v},w)\overline{u}+(\psi^{-1}R)(w,\overline{u})\overline{v} &= \\ \varphi^{-1}\left(R(\overline{v},w)\right)\overline{u}+\varphi^{-1}\left(\overline{R(\overline{w},u)}\right)\overline{v} &= \\ -\varphi^{-1}\left(\overline{R(\overline{w},v)}\right)\overline{u}+\varphi^{-1}\left(\overline{R(\overline{w},u)}\right)\overline{v} &= \\ -\overline{R(\overline{w},v)u}+\overline{R(\overline{w},u)v} &= 0\end{aligned}$$

Terms with entries only from V or only from $\overline{V}$ are zero. □

In contrary to the previous proof, in this proof we only supposed the fact that θ^h is non-degenerate and not that $h^{\mathbb{C}}_{|V\times V} = 0$. If we assume h to be positive definite, then both facts are satisfied.

4.4.2 Consequences for Berger and weak-Berger algebras

Both propositions give three important corollaries.

4.56 Corollary. *Let $\mathfrak{h}_0 \subset \mathfrak{g}_0 \subset \mathfrak{so}(E^{\mathbb{C}}, H)$ be subalgebras of non-real type, $\mathfrak{h}$ and $\mathfrak{g}$ defined as above. If*

$$\mathfrak{h}^{(1)} = \mathfrak{g}^{(1)},$$

then $(\mathfrak{h}_0^{\mathbb{C}})_H = (\mathfrak{g}_0^{\mathbb{C}})_H$. I.e. if in $\mathfrak{g}$ exists a proper subalgebra which has the same first prolongation and a compact real form in $\mathfrak{g}_0$ of non-real type, then $\mathfrak{g}_0^{\mathbb{C}}$ and therefore $\mathfrak{g}_0$ cannot be weak-Berger algebras.

Proof. Because of $Q \in \mathcal{B}_H(\mathfrak{h}_0^{\mathbb{C}}) \simeq \mathfrak{h}^{(1)} = \mathfrak{g}^{(1)} \simeq \mathcal{B}_H(\mathfrak{g}_0^{\mathbb{C}})$ we have $Q(u) \in (\mathfrak{g}_0^{\mathbb{C}})_H$ if and only if $Q(u) \in (\mathfrak{h}_0^{\mathbb{C}})_H$. □

4.57 Corollary. *Let $\mathfrak{g}_0 \subset \mathfrak{so}(E^{\mathbb{C}}, H)$ be a Lie algebra of non-real type, and $\mathfrak{g}$ defined as above. Then*

1. *$(\mathfrak{g}_0^{\mathbb{C}})_H = \mathfrak{g}_0^{\mathbb{C}}$ (i.e. $\mathfrak{g}_0^{\mathbb{C}}$ is a weak-Berger-algebra) if and only if $\mathfrak{g} = \tilde{\mathfrak{g}}$.*
2. *$\underline{\mathfrak{g}_0^{\mathbb{C}}} = \mathfrak{g}_0^{\mathbb{C}}$ (i.e. $\mathfrak{g}_0^{\mathbb{C}}$ is a Berger-algebra) if and only if $\mathfrak{g} = \tilde{\tilde{\mathfrak{g}}}$.*

Proof. 1.) We show first the sufficiency: Let $A \in \mathfrak{g}_0^{\mathbb{C}}$ be arbitrary. The assumption $\mathfrak{g} = \tilde{\mathfrak{g}}$ gives w.l.o.g. that $\varphi(A) = Q(u)$ with $Q \in \mathfrak{g}^{(1)}$ and $u \in V$. But then we have

$$(\phi^{-1}Q)(u) \overset{\text{per def.}}{=} \varphi^{-1}(Q(u)) = \varphi^{-1}(\varphi(A)) = A,$$

with $(\phi^{-1}Q) \in \mathcal{B}_H(\mathfrak{g}_0^{\mathbb{C}})$, i.e. $A \in (\mathfrak{g}_0^{\mathbb{C}})_H$.

Now we show the necessity: If $A \in \mathfrak{g}$, then the assumption $\mathfrak{g}_0^{\mathbb{C}} = (\mathfrak{g}_0^{\mathbb{C}})_H$ gives w.l.o.g. that $\varphi^{-1}(A) = \hat{Q}(u + \overline{v})$ with $\hat{Q} \in \mathcal{B}_H(\mathfrak{g}_0^{\mathbb{C}})$, $u \in V$ and $\overline{v} \in \overline{V}$. But by the isomorphism of the proposition 4.54 there is a $Q \in \mathfrak{g}^{(1)}$ such that

$$\varphi^{-1}(A) = \hat{Q}(u + \overline{v}) = (\phi^{-1}Q)(u + \overline{v}) = \varphi^{-1}(Q(u)) + \varphi^{-1}(\overline{Q(v)}).$$

But this means that

$$A \;=\; \underbrace{Q(u)}_{\in \tilde{\mathfrak{g}}} + \underbrace{\overline{Q(v)}}_{\in \tilde{\mathfrak{g}}} \;\in\; \tilde{\mathfrak{g}},$$

i.e. $\mathfrak{g} \subset \tilde{\mathfrak{g}}$.
2.) Both directions are proved completely analogous to 1.)
Suppose that $\mathfrak{g} = \tilde{\tilde{g}}$. Then for $A \in \mathfrak{g}_0^{\mathbb{C}}$ one has that $\varphi(A) = R(\overline{u}, v)$ and

$$(\psi^{-1}R)(\overline{u}, v) \;=\; \varphi^{-1}(R(\overline{u}, v) \;=\; A.$$

On the other hand we have for $A \in \mathfrak{g}$ that $\varphi^{-1}(A) = \hat{R}(z + \overline{u}, v + \overline{w})$. This gives

$$\begin{aligned} \varphi^{-1}(A) &= \hat{R}(z, \overline{w}) + \hat{R}(\overline{u}, v) &&= (\psi^{-1}R)(z, \overline{w}) + = (\psi^{-1}R)(\overline{u}, v) \\ &= \varphi^{-1}(\overline{R(\overline{z}, w)}) + \varphi^{-1}(R(\overline{u}, v)) && . \end{aligned}$$

and therefore $A \in \tilde{\tilde{\mathfrak{g}}}$. □

For the next corollary we consider a special case, the relevance of which will be clearer in the appendix C, where we prove the main proposition, after the classification of non-vanishing first prolongations.

4.58 Corollary. *Let be $\mathfrak{g} \subset \mathfrak{gl}(V)$ with the same notations as above. We suppose that there is an isomorphism between*

$$\begin{aligned} V^* &\simeq \mathfrak{g}^{(1)} \quad \textit{as vector spaces, not as modules} \\ \sigma &\longmapsto Q_\sigma. \end{aligned}$$

Then the hermitian form θ defines an isomorphism (not of modules, of course)

$$\begin{aligned} \beta \;:\; \overline{V} &\simeq \mathfrak{g}^{(1)} \\ \overline{u} &\longmapsto \beta(\overline{u}, .) := Q_{\theta(.,u)}, \quad \textit{i.e. } \beta(\overline{u}, v) := Q_{\theta(.,u)} v \in \mathfrak{g}. \end{aligned}$$

If furthermore $\beta \in \mathfrak{g}^{[1,1]} \subset \overline{V}^* \otimes \mathfrak{g}^{(1)} \simeq Hom(\overline{V}, \mathfrak{g}^{(1)})$, i.e. if the condition

$$\overline{\beta(\overline{u}, v)} \;=\; -\beta(\overline{v}, u) \tag{4.60}$$

is fulfilled, then $\tilde{\mathfrak{g}} \subset \tilde{\tilde{g}}$, i.e. if $\mathfrak{g}_0^{\mathbb{C}}$ is a weak-Berger-algebra, then it is a Berger- algebra.

The *Proof* is obvious: The existence of the isomorphism is clear since θ identifies V^* and $\overline{V}$.
If $Q(v) \in \tilde{\mathfrak{g}}$, then by the assumption $Q = \beta(\overline{u}, .)$, i.e. $Q(v) = \beta(\overline{u}, v)$. If now β fulfills the condition with respect to the conjugation in $\mathfrak{g}$, then we have $\beta \in \mathfrak{g}^{[1,1]}$ and thus $\beta(\overline{u}, v) \in \tilde{\tilde{\mathfrak{g}}}$. □

As a result of the previous and this section we have to investigate complex irreducible representations of complex Lie algebras with non-vanishing first prolongation. Fortunately these are classified by Cartan [Car09], Kobayashi and Nagano [KN65] in a rather short list. In the next section we will present this list and check for the entries with the help of the previous corollaries whether they are Berger or weak-Berger algebras.

4.4.3 Lie algebras with non-trivial first prolongation

There are only a few complex Lie algebras $\mathfrak{g}$ contained irreducibly in $\mathfrak{gl}(V)$which have non vanishing first prolongation. The classification is due to [Car09] and [KN65]. We will cite them following [MS99] in two tables.

Table 1 Complex Lie-groups and algebras with $\mathfrak{g}^{(1)} \neq 0$ and $\mathfrak{g}^{(1)} \neq V^*$:

	G	$\mathfrak{g}$	V		$\mathfrak{g}^{(1)}$
1.	$Sl(n, \mathbb{C})$	$\mathfrak{sl}(n, \mathbb{C})$	$\mathbb{C}^n$,	$n \geq 2$	$(V \otimes \odot^2 V^*)_0$
2.	$Gl(n, \mathbb{C})$	$\mathfrak{gl}(n, \mathbb{C})$	$\mathbb{C}^n$,	$n \geq 1$	$V \otimes \odot^2 V^*$
3.	$Sp(n, \mathbb{C})$	$\mathfrak{sp}(n, \mathbb{C})$	$\mathbb{C}^{2n}$,	$n \geq 2$	$\odot^3 V^*$
4.	$\mathbb{C}^* \times Sp(n, \mathbb{C})$	$\mathbb{C} \oplus \mathfrak{sp}(n, \mathbb{C})$,	$\mathbb{C}^{2n}$,	$n \geq 2$	$\odot^3 V^*$

Table 2 Complex Lie-groups and algebras with first prolongation $\mathfrak{g}^{(1)} = V^*$:

	G	$\mathfrak{g}$	V	
1.	$CO(n, \mathbb{C})$	$\mathfrak{co}(n, \mathbb{C})$	$\mathbb{C}^n$,	$n \geq 3$
2.	$Gl(n, \mathbb{C})$	$\mathfrak{gl}(n, \mathbb{C})$	$\odot^2 \mathbb{C}^n$,	$n \geq 2$
3.	$Gl(n, \mathbb{C})$	$\mathfrak{gl}(n, \mathbb{C})$	$\Lambda^2 \mathbb{C}^n$,	$n \geq 5$
4.	$Gl(n, \mathbb{C}) \cdot Gl(m, \mathbb{C})$	$\mathfrak{sl}(\mathfrak{gl}(n, \mathbb{C}) \oplus \mathfrak{gl}(m, \mathbb{C}))$	$\mathbb{C}^n \otimes \mathbb{C}^m$,	$m, n \geq 2$
5.	$\mathbb{C}^* \cdot Spin(10, \mathbb{C})$	$\mathbb{C} \oplus \mathfrak{spin}(10, \mathbb{C})$	$\Delta_{10}^+ \simeq \mathbb{C}^{16}$	
6.	$\mathbb{C}^* \cdot E_6$	$\mathbb{C} \oplus \mathfrak{e}_6$	$\mathbb{C}^{27}$	

We have to make two remarks about the second table:
The fourth Lie algebra is defined as

$$\begin{aligned} \mathfrak{sl}(\mathfrak{gl}(n, \mathbb{C}) \oplus \mathfrak{gl}(m, \mathbb{C})) &= \{(X, Y) \in \mathfrak{gl}(n, \mathbb{C}) \oplus \mathfrak{gl}(m, \mathbb{C}) | tr\ X + tr\ Y = 0\} \\ &= (\mathfrak{gl}(n, \mathbb{C}) \oplus \mathfrak{gl}(m, \mathbb{C})) \cap \mathfrak{sl}(n + m, \mathbb{C}). \end{aligned}$$

The identification with the Lie algebra of the group is given as follows

$$\begin{array}{rcl} \mathfrak{sl}(\mathfrak{gl}(n,\mathbb{C}) \oplus \mathfrak{gl}(m,\mathbb{C})) & \simeq & LA(Gl(n,\mathbb{C}) \cdot GL(m,\mathbb{C})) \subset \mathfrak{gl}(n \cdot m,\mathbb{C}) \\ (A,B) & \longmapsto & (x \otimes u \mapsto Ax \otimes u - x \otimes Bu). \end{array}$$

In entry 5. Δ_{10}^{+} denotes the irreducible $Spin(10,\mathbb{C})$ spinor module. The representation in 6. is one of the two 27-dimensional, irreducible $\mathfrak{e}_6$ representations, which are conjugate to each other as representations of the compact real form of $\mathfrak{e}_6$. We will explain these representations in the appendix C.

The algebras of table 1 The first three entries of table 1 are all complexifications of Riemannian holonomy algebras $\mathfrak{su}(n)$, $\mathfrak{u}(n)$ acting on $\mathbb{R}^{2n}$ and $\mathfrak{sp}(n)$ acting on $\mathbb{R}^{4n}$ and therefore Berger algebras.
The fourth has the compact real form $i\mathbb{R} \oplus \mathfrak{sp}(n) \simeq \mathfrak{so}(2) \oplus \mathfrak{sp}(n)$ acting irreducible on $\mathbb{R}^{4n}$ where $i\ id$ corresponds to the element $J \in \mathfrak{u}(2n)$. Since the representation of $\mathfrak{sp}(n)$ on $\mathbb{R}^{4n}$ is of non-real type we are in the situation of corollary 4.56, because $(\mathbb{C}Id \oplus \mathfrak{sp}(n,\mathbb{C}))^{(1)} = \mathfrak{sp}(n,\mathbb{C})^{(1)}$. Hence $\mathbb{C} \oplus sp(2n,\mathbb{C})$ is not a weak-Berger algebra.

The algebras of table 2 If one looks at the unique compact real form and the reellification of the Lie algebras and representations in table 2 — we will do this in extenso in the appendix — one sees that they correspond to the holonomy representation of Riemannian symmetric spaces which are Kählerian. This gives the following proposition, which will be verified in the appendix.

4.59 Proposition. *The compact real forms of the algebras in table 2 and the reellification of the representations are equivalent to the holonomy representations of the following Riemannian, Kählerian symmetric spaces (see [Hel78]):*

	Type	*non-compact*	*compact*	*dim.*
1.	*BD I*	$SO_0(2,n)/SO(2) \times SO(n)$	$SO(2+n)/SO(2) \times SO(n)$	$2n$
2.	*C I*	$Sp(n,\mathbb{R})/U(n)$	$Sp(n)/U(n)$	$n(n+1)$
3.	*D III*	$SO^*(2n)/U(n)$	$SO(2n)/U(n)$	$n(n-1)$
4.	*A III*	$SU(n,m)/U(n) \cdot U(m)$	$SU(n+m)/U(n) \cdot U(m)$	$2nm$
5.	*E III*	$(\mathfrak{e}_{6(-14)}, \mathfrak{so}(2) \oplus \mathfrak{so}(10))$	$(\mathfrak{e}_{6(-78)}, \mathfrak{so}(2) \oplus \mathfrak{so}(10))$	32
6.	*E VII*	$(\mathfrak{e}_{7(-25)}, \mathfrak{so}(2) \oplus \mathfrak{e}_6)$	$(\mathfrak{e}_{7(-133)}, \mathfrak{so}(2) \oplus \mathfrak{e}_6)$	54

Table 3 *Riemannian, Kählerian symmetric spaces corresponding to table 2*

4.4.4 The result and consequences

The conclusion from the previous section is: every real Lie algebra $\mathfrak{g}_0$ of non-real type, i.e. contained in $\mathfrak{u}(n)$, that can be weak-Berger is a Berger algebra. Further each of these Lie algebras is the holonomy algebra of a Riemannian manifold, the remaining entries of table 1 of non-symmetric ones, and the entries of table 2 of symmetric ones.

4.60 Theorem. *Let $\mathfrak{g}$ be a Lie algebra and E an irreducible $\mathfrak{g}$–module of non-real type. If $\mathfrak{g} \subset \mathfrak{so}(E,h)$ is a weak-Berger algebra then it is a Berger algebra, in particular a Riemannian holonomy algebra.*

Before we we apply this to the irreducible components of the $\mathfrak{so}(n)$-projection of the holonomy algebra of an indecomposable Lorentzian manifold with light like invariant subspace, we prove a lemma to get the result in full generality.

4.61 Lemma. *Let $\mathfrak{g} \subset \mathfrak{u}(n) \subset \mathfrak{so}(2n)$ be a Lie algebra with the decomposition property of theorem 2.7, ie. there exists decompositions of $\mathbb{R}^{2n}$ into orthogonal subspaces and of $\mathfrak{g}$ into ideals*

$$\mathbb{R}^{2n} = E_0 \oplus E_1 \oplus \ldots \oplus E_r \quad \text{and} \quad \mathfrak{g} = \mathfrak{g}_1 \oplus \ldots \oplus \mathfrak{g}_r$$

where $\mathfrak{g}$ acts trivial on E_0, $\mathfrak{g}_i$ acts irreducible on E_i and $\mathfrak{g}_i(E_j) = \{0\}$ for $i \neq j$. Then $\mathfrak{g} \subset \mathfrak{u}(n)$ implies dim $E_i = 2k_i$ and $\mathfrak{g}_i \subset \mathfrak{u}(k_i)$ for $i = 1, \ldots, r$.

Proof. Let $\mathbb{R}^{2n} = \mathbb{C}^n$ and θ be the positive definite hermitian form on $\mathbb{C}^n$. Let E_i be an invariant subspace on which $\mathfrak{g}$ acts irreducible. If $E_i = V^i_{\mathbb{R}}$ for a complex vector space V^i, then we can restrict θ to V^i. Because θ is positive definite it is non-degenerate on V^i — since $\theta(v,v) > 0$ for $v \neq 0$ — we get that $\mathfrak{g}_i \subset \mathfrak{u}(V^i, \theta)$, i.e. $\mathfrak{g} \subset \mathfrak{u}(k_i)$.
Hence we have to consider a subspace E_i which is not the reellification of a complex vector space. Let J be the complex structure on $\mathbb{R}^{2n}$. We consider the real vector space JE_i, which is invariant under $\mathfrak{g}$, since J commutes with $\mathfrak{g}$. Then the space $JE_i \cap E_i$ is contained in E_i as well as in JE_i and invariant under $\mathfrak{g}$. Because $\mathfrak{g}$ acts irreducible on E_i we get two cases. The first is $E_i \cap JE_i = E_i = JE_i$, but this was excluded since E_i was not a reellification. The second is $E_i \cap JE_i = \{0\}$. So we have two invariant irreducible subspaces on which $\mathfrak{g}$ acts simultaneously, i.e. $A(x, Jy) = (Ax, AJy)$, but this is not possible because of the Borel-Lichnerowicz decomposition property from theorem 2.7. □

4.62 Theorem. *Let (M,h) be an indecomposable $n+2$-dimensional Lorentzian manifold with light like holonomy-invariant subspace and set $\mathfrak{g} := pr_{\mathfrak{so}(n)}\mathfrak{hol}_p(M,h)$. Then holds:*

1. *Any irreducible component $\mathfrak{g}_i$ of $\mathfrak{g}$ (due to theorem 2.7) which is unitary, i.e. $\mathfrak{g}_i \subset \mathfrak{u}(d_i/2)$, for d_i the dimension of E_i, is the holonomy algebra of a Riemannian manifold.*

2. *If $\mathfrak{g} \subset \mathfrak{u}(n) \subset \mathfrak{so}(2n)$. Then $\mathfrak{g}$ is the holonomy algebra of a Riemannian manifold.*

Proof. $\mathfrak{g} \subset \mathfrak{so}(n)$ is a weak-Berger algebra. Then all the $\mathfrak{g}_i \subset \mathfrak{so}(d_i)$ of the decomposition of theorem 2.7 are weak-Berger because of corollary 4.5. To those of them which are unitary, i.e.$\mathfrak{g}_i \subset \mathfrak{u}(d_i/2)$ theorem 4.60 applies and give point 1.
For the second point $\mathfrak{g} \subset \mathfrak{u}(n/2)$ implies by the lemma that $\mathfrak{g}_i \subset \mathfrak{u}(d_i/2)$. Again theorem 4.60 says that each $\mathfrak{g}_i$ and hence $\mathfrak{g}$ is a Riemannian holonomy algebra. □

4.63 Remark. Quaternionic symmetric spaces. Again we have to make a remark about quaternionic symmetric spaces (see remark 4.52). If $G/Sp(1) \cdot H$ with $H \subset Sp(n)$ is a quaternionic symmetric space then the corresponding complex irreducible representation of H is of quaternionic, i.e. of non real type, and it is Berger [Sch99]. But the real representation of H, i.e. the reellification of the complex one, is not. Thats why it does not occur in the above list. The place of $Sp(1)\mathfrak{d}otH$ would be in a list of real semisimple, but non-simple, weak-Berger algebras of real type.

Finally we will formulate a corollary drawing consequences for the existence of Lorentzian manifolds with holonomy of the coupled types 3 and 4 due to theorem 2.10. For the coupled type the $\mathfrak{so}(n)$–projection has to have a non-trivial center, i.e. one of its irreducibly acting ideals due to theorem 2.7 has to have a non-trivial center. Hence this component is contained in $\mathfrak{u}(d)$. Then it has to be an irreducible Riemannian holonomy algebra with center. So we get the following possibilities.

4.64 Corollary. *Let (M, h) be an indecomposable, non-irreducible Lorentzian manifold with holonomy of coupled type 3 or 4, and let $\mathfrak{g}$ be its $\mathfrak{so}(n)$–projection. Then at least one of its irreducibly acting ideals of $\mathfrak{g}$ is a Riemannian holonomy algebra with center, i.e. it equals to $\mathfrak{u}(d) \subset \mathfrak{so}(2d)$ or to the holonomy algebra of a Riemannian hermitian symmetric space listed in table 3.*

This leads to a consequence for the existence of parallel spinors on manifolds with holonomy of coupled type.

4.65 Corollary. *If an indecomposable Lorentzian manifold carries a parallel spinor, then its reduced holonomy group is of uncoupled type $G \ltimes \mathbb{R}^n$.*

Proof. Consider an $n+2$–dimensional indecomposable Lorentzian manifold with parallel spinor. On one hand indecomposability ensures that this spinor induces a lightlike parallel vector field. Hence the manifold is non-irreducible with holonomy conatained in $SO(n) \ltimes \mathbb{R}^n$, i.e. of type 2 or 4. One the other hand the existence of a parallel spinor implies by proposition 3.1 and corollary 3.13 that each of the irreducible components has a fixed spinor under its spin representation. But this is neither possible for $\mathfrak{u}(d) \subset \mathfrak{so}(2d)$ nor for the holonomy algebra of a Riemannian hermitian symmetric space. Hence the previous corollary implies that the holonomy of the manifold cannot be of coupled type 4. □

Appendix A

Complex semi-simple Lie algebras and their representations

In this appendix we will recall the main parts of the theory of semisimple complex Lie algebras and their irreducible representations.

A.1 Structure of a complex semi-simple Lie algebra

Cartan subalgebra and root decomposition Let $\mathfrak{g}$ be a (complex or real) Lie algebra and $\mathfrak{t}$ the Cartan subalgebra, i. e. $\mathfrak{t}$ is nilpotent and its own normalizer. A Cartan subalgebra always exists and for complex Lie algebras it induces a root space decomposition of $\mathfrak{g}$, i. e. there exist $\alpha_i \in \mathfrak{h}^*$ such that

$$\mathfrak{g} = \mathfrak{g}_{\alpha_1} \oplus \ldots \oplus \mathfrak{g}_{\alpha_r}$$

with $\mathfrak{g}_{\alpha_i} := \{Y \in \mathfrak{g} |$ for each $X \in \mathfrak{t}$ there is a $k > 0$ such that $(adX - \alpha_i(X))\, Y = 0\}$. From the defining properties of a Cartan algebra follows $\mathfrak{t} = \mathfrak{g}_0$.

For complex semisimple Lie algebras the Cartan subalgebra is maximal abelian and the Killing form B is non degenerate on $\mathfrak{t}$. Furthermore $ad(H) : \mathfrak{g} \to \mathfrak{g}$ is diagonizable for every $H \in \mathfrak{t}$. So we have the root space decomposition in $\mathfrak{g}(\alpha_i) := \{Y \in \mathfrak{g} |$ for each $X \in \mathfrak{t}$ it is $adX(Y) = \alpha_i(X)\ Y\}$ and

$$\mathfrak{g} = \mathfrak{t} \oplus \mathfrak{g}_{\alpha_1} \oplus \ldots \oplus \mathfrak{g}_{\alpha_r}.$$

We denote by

$$\Delta := \{\alpha \in \mathfrak{t}^* | \mathfrak{g}_\alpha \neq 0\}$$

the roots of $\mathfrak{g}$. For every $\alpha \in \Delta$ holds

$$dim\ \mathfrak{g}_\alpha = dim\ [\mathfrak{g}_\alpha, \mathfrak{g}_{-\alpha}] = 1.$$

Example $\mathfrak{sl}(2,\mathbb{C})$ We consider the the simple complex Lie algebra $\mathfrak{sl}(2,\mathbb{C})$ spanned by the following matrices

$$H = \begin{pmatrix} 1 & 0 \\ 0 & -1 \end{pmatrix},\ X = \begin{pmatrix} 0 & 1 \\ 0 & 0 \end{pmatrix},\ Y = \begin{pmatrix} 0 & 0 \\ 1 & 0 \end{pmatrix}$$

The commutator relations are the following

$$[X,Y] = H,\ [H,X] = 2X,\ [H,Y] = -2Y.$$

$\mathfrak{sl}(2,\mathbb{C})$ has two roots:

$$\begin{array}{llll} \alpha_+ : H \mapsto 2 & , & \alpha_- H \mapsto -2 & \text{with} \\ \mathfrak{g}_{\alpha_+} = \mathbb{C}X & , & \mathfrak{g}_{\alpha_-} = \mathbb{C}Y & \text{and } \mathfrak{t} = \mathbb{C}H \end{array}$$

Dual vectors For every $\alpha \in \Delta$ one defines the dual vector h_α by $\alpha(H) = B(h_\alpha, H)$. Then it is for $X \in \mathfrak{g}_\alpha, Y \in \mathfrak{g}_{-\alpha}$:

$$[X,Y] = B(X,Y)h_\alpha.$$

Since B is non degenerate on $\mathfrak{t}$ we can normalize the dual vectors:

$$H_\alpha := \frac{2}{B(h_\alpha, h_\alpha)} h_\alpha$$

such that $\alpha(H_\alpha) = 2$. So we have for every $\alpha \in \Delta$ a $H_\alpha \in \mathfrak{t}$ and can choose an $X_\alpha \in \mathfrak{g}_\alpha$ and an $X_{-\alpha} \in \mathfrak{g}_{-\alpha}$ and such that

$$[X_\alpha, X_{-\alpha}] = H_\alpha,\ [H_\alpha, X_\alpha] = 2X_\alpha,\ [H_\alpha, X_{-\alpha}] = -2X_{-\alpha},$$

i. e. $span(H_\alpha, X_\alpha, X_{-\alpha})$ is isomorphic to $\mathfrak{sl}(2,\mathbb{C})$.
Furthermore the following properties are valid:

1. $\mathfrak{t} = span(H_\alpha | \alpha \in \Delta)$,
2. $\beta(H_\alpha) \in \mathbb{Z}$,

Euclidian vector spaces Now one defines real subspaces of $\mathfrak{t}$ resp. $\mathfrak{t}^*$:

$$\begin{aligned} \mathfrak{t}_0 &:= span_{\mathbb{R}}(H_\alpha | \alpha \in \Delta) = span_{\mathbb{R}}(h_\alpha | \alpha \in \Delta) \\ \mathfrak{t}_0^* &:= span_{\mathbb{R}}(\alpha | \alpha \in \Delta). \end{aligned}$$

Then the Killing form is positive definite on $\mathfrak{t}_0$. Furthermore $\langle \alpha, \beta \rangle := B(h_\alpha, h_\beta)$ is a positive bilinear form on $\mathfrak{t}_0^*$. Hence $\mathfrak{t}_0$ and $\mathfrak{t}_0^*$ turn into euclidian vector spaces.

Positive, negative and fundamental roots Every root system decomposes in positive and negative roots $\Delta = \Delta_+ \cup \Delta_-$ and one defines a fundamental root system Σ by

$$\Sigma := \{\alpha \in \Delta_+ | \alpha \text{ is not the sum of two positive roots}\}.$$

Then we have for every root $\Delta_\pm \ni \beta = \pm \sum_{a_i \in \Sigma} b_i \alpha_i$ with $b_i \in \mathbb{N}$.
For $\mathfrak{g}$ we have now the decomposition

$$\mathfrak{g} = \mathfrak{t} \oplus \oplus_{\alpha \in \Delta_+} (\mathfrak{g}_\alpha \oplus \mathfrak{g}_{-\alpha}).$$

$\mathfrak{n}_\pm := \oplus_{\alpha \in \Delta_+} \mathfrak{g}(\pm\alpha)$ are nilpotent and $\mathfrak{b} := \mathfrak{t} \oplus_{\alpha \in \Delta_+} \mathfrak{g}_\alpha$ is solvable and called Borel algebra.

The Weyl basis On defines the so called Weyl-basis $(X_\alpha \in \mathfrak{g}_\alpha | \alpha \in \Delta)$ of $\mathfrak{g}$ *mod* $\mathfrak{t}$ defined by the following properties

$$\begin{aligned} [H, X_\alpha] &= \alpha(H) X_\alpha \text{ for all } H \in \mathfrak{h} \\ [X_\alpha, X_{-\alpha}] &= h_\alpha \\ [X_\alpha, X_\beta] &= N_{\alpha,\beta} X_{\alpha+\beta} \end{aligned}$$

with $N_{\alpha,\beta} = 0$ if $\alpha + \beta \notin \Delta$ and $N_{\alpha,\beta} = N_{-\alpha,-\beta} = -N_{\beta,\alpha}$ and some additional properties.

The Weyl group On the euclidian vector space $\mathfrak{t}_0^*$ we have the Weyl group W acting. This group is generated by the following reflections

$$s_\alpha : \beta \mapsto \beta - 2\frac{\langle \beta, \alpha \rangle}{\langle \alpha, \alpha \rangle} \alpha \text{ for } \alpha \in \Delta.$$

W is finite and we have $W\Delta = \Delta$. There exists an element $\sigma \in W$ such that $\sigma(\Delta_+) = \Delta_-$. Furthermore W acts simply transitive on the fundamental root systems.

Compact real forms A real Lie algebra $\mathfrak{g}$ is called compact if the group of inner automorphisms is a compact subgroup of the automorphism group of $\mathfrak{g}$.
Compact Lie algebras are reductive, i.e. the Levi decomposition is given by

$$\mathfrak{g} = \mathfrak{z}(\mathfrak{g}) \oplus [\mathfrak{g}, \mathfrak{g}] \text{ where } \mathfrak{z}(\mathfrak{g}) \text{ the center of } \mathfrak{g}.$$

If $\mathfrak{g}$ is semisimple the compactness is equivalent to the fact that the Killing form is negative definite.
Every semisimple complex Lie algebra $\mathfrak{g}$ has a compact real form $\mathfrak{g}_0$. It is given by means of the Weyl-basis X_α with respect to a Cartan subalgebra $\mathfrak{t}$ of $\mathfrak{g}$:

Set $\mathfrak{t}_0 := span_{\mathbb{R}}(H_\alpha|\alpha \in \Delta)$. Then $\mathfrak{t} = \mathfrak{t}_0^{\mathbb{C}}$

$$\mathfrak{g}_0 := i\ \mathfrak{t}_0 \oplus span_{\mathbb{R}}\left(X_\alpha - X_{-\alpha}, i\left(X_\alpha + X_{-\alpha}\right)|\alpha \in \Delta_+\right). \tag{A.1}$$

Let $\mathfrak{g}_0$ be a compact real form of a complex semisimple Lie algebra, corresponding to the real structure A. Define $\mathfrak{k}_A := \{X \in \mathfrak{g}_0 | AX = X\}$ and $\mathfrak{p}_A := \{X \in \mathfrak{g}_0 | AX = -X\}$. Then

$$\mathfrak{g}_0 = \mathfrak{k}_A \oplus i\ \mathfrak{p}_A.$$

This decomposition is called Cartan decomposition of $\mathfrak{g}$.
The compact real form of a complex semisimple Lie algebra $\mathfrak{g}$ is uniquely defined up to inner autmorphisms of $\mathfrak{g}$.

A.2 Basic facts about complex representations

Let now (κ, V) be an complex, irreducible, finite dimensional representation of a complex semisimple Lie algebra $\mathfrak{g}$. For $\lambda \in \mathfrak{t}^*$ one defines the weight spaces

$$V_\lambda := \{v \in V | \kappa(H)v = \lambda(H)v \text{ for all } H \in \mathfrak{t}\}$$

and the weights

$$\Omega := \{\lambda \in \mathfrak{t}^* | V_\lambda \neq 0\}$$

Then V decomposes as follows $V = \oplus_{\lambda\in\Omega} V_\lambda$. An important relation is

$$\kappa(\mathfrak{g}_\alpha)V_\lambda \subset V_{\lambda+\alpha}.$$

Primitive vector and highest weight With respect to a Borel algebra $\mathfrak{b} = \mathfrak{t} \oplus \mathfrak{n}_+$ we have a unique primitive vector $v \in V(\Lambda)$ of weight Λ, i.e. $\kappa(\mathfrak{n}_+)v = 0$. Λ is called highest weight (because $\Lambda + \alpha$ in no weight for every $\alpha \in \Delta_+$). It is also unique and has multiplicity 1.
For every $\Lambda \in \mathfrak{t}^*$ there is a irreducible representation κ with highest weight Λ and two representations of the same highest weight are isomorphic.

Weights and roots For $a \in \Delta_+$ one considers the elements $Y_\alpha := X_{-\alpha} \in \mathfrak{n}_-$. Then we have that V is generated by

$$\kappa\left(Y_{\alpha_1}^{p_1} \cdot \ldots \cdot Y_{\alpha_k}^{p_k}\right) v \in V(\Lambda - \sum_{i=1}^{k} p_i\alpha_i) \text{ with } k > 0 \text{ and } \alpha_i \in \Delta_+$$

Furthermore every weight has the form $\Lambda - \sum_{i=1}^{k} q_i\pi_i$ with $\pi_i \in \Sigma$ fundamental roots and $q_i \in \mathbb{N}$.

Fundamental weights Let $\Sigma = (\pi_1, \dots, \pi_k)$ be a fundamental root system, $H_i := H_{\pi_i} \in \mathfrak{t}_0^*$. The fundamental weights $(\omega_1, \dots, \omega_k)$ are defined by the relation $\omega_i(H_j) = \delta_{ij}$. The representations with highest weight ω_i is called fundamental representation of $\mathfrak{g}$. The highest weight Λ of a representation κ can now be written as

$$\Lambda = \sum_{i=1}^{k} m_i\, \omega_i \text{ with } m_i \in \mathbb{N}.$$

The dimension of an irreducible representation One defines $\delta := \frac{1}{2} \sum_{\alpha \in \Delta_+} \alpha$. It is $\delta(H_\alpha) \in \mathbb{Z}$ for all $\alpha \in \Delta$. It is clear that $\delta(H_i) = 1$ for all $i = 1, \dots, k$ such that $\delta = \omega_1 + \ldots + \omega_k$.

Let (κ, V) be a representation of highest weight Λ. Then for the dimension of V holds

$$dim\ V = \prod_{\alpha \in \Delta_+} \frac{(\Lambda + \delta)(H_\alpha)}{\delta(H_\alpha)} = \prod_{\alpha \in \Delta_+} \frac{((m_1 + 1)\omega_1 + \ldots + (m_k + 1)\omega_k)(H_\alpha)}{(\omega_1 + \ldots + \omega_k)(H_\alpha)}.$$

The dual representation Let (κ, V) be an irreducible representation of $\mathfrak{g}$. One defines the dual representation κ^* on V^* as follows

$$\kappa^*(X) : \nu \in V^* \mapsto (v \mapsto -\nu(\kappa(X)v)).$$

Let σ the element of the Weyl group for which $\sigma(\Delta_+) = \Delta_-$. If κ is of highest weight Λ then κ^* is of highest weight $-\sigma(\Lambda)$.

Since the existence of an invariant bilinear form for κ is equivalent to the self duality of κ we have

A.1 Proposition. *[Bou75][7, nr. 5] Let (κ, V) be an complex irreducible representation of a complex semi-simple Lie algebra of highest weight Λ and σ as above. Then exists an invariant bilinear form for κ if and only if $\sigma(\Lambda) = -\Lambda$. This bilinear form is unique and it is symmetric/skew symmetric if and only if $\sum_{\alpha \in \Delta_+} \Lambda(H_\alpha)$ is even/odd.*

The proof uses the representations of $\mathfrak{sl}(2, \mathbb{C})$. Although we will not gives this proof here we will describe the irreducible $\mathfrak{sl}(2, \mathbb{C})$-representations.

Example $\mathfrak{sl}(2, \mathbb{C})$ The representations $\pi_m : \mathfrak{sl}(2, \mathbb{C}) \to \mathfrak{gl}(\mathbb{C}^{m+1})$ defined by the following action on the canonical basis $v_0, \dots, v_m$ of $\mathbb{C}^{m+1}$:

$$\begin{aligned} \pi_m(H)(v_k) &= (m - 2k)v_k \quad \text{for } k = 0, \dots, m \\ \pi_m(Y)(v_k) &= (k+1)v_{k+1} \quad \text{for } k = 0, \dots, m-1 \text{ and } \pi_m(Y)(v_m) = 0 \\ \pi_m(X)(v_k) &= (m - (k-1))\, v_{k-1} \quad \text{for } k = 1, \dots, m \text{ and } \pi_m(X)(v_0) = 0 \end{aligned}$$

These representations are irreducible in every dimension and every irreducible representation of $\mathfrak{sl}(2, \mathbb{C})$ in dimension $m+1$ is equivalent to π_m.

It is $v_k \in V(m-2k)$ and the highest weight is $\Lambda(H) = m$.
One defines an 2-form β on $\mathbb{C}^{m+1}$ by the formula

$$\beta(v_k, v_{m-k}) := (-1)^k \binom{m}{k}, \quad \text{for } k = 0, \ldots, m \tag{A.2}$$

An easy calculation shows that β is $\mathfrak{sl}(2,\mathbb{C})$ invariant, i.e. the π_m are self-dual. This is clear because the element σ of the Weyl group, mapping $\Delta_+ = \{\alpha_+\}$ on $\Delta_- = \{\alpha_-\}$ is -1.
Furthermore one gets that

$$\beta(v_{m-k}, v_k) = (-1)^{m+k} \binom{m}{k} = (-1)^m \beta(v_k, v_{m-k})$$

i.e. β is symmetric if m is even and skew if m is odd. That means the π_m are orthogonal representations for m even and symplectic ones for m odd.

Appendix B

Representations of real Lie algebras

In this appendix we will collect and illustrate some standard facts about representations of real Lie algebras.
Because of the theorem 2.10 and proposition 4.5 we are interested in irreducible real representations of real Lie algebras which are orthogonal.
First we will recall some facts about irreducible complex representations of real Lie algebras, in particular orthogonal or unitary ones.
Then we will use the results of E. Cartan ([Car14], see also [Got78], pp.363 and [Iwa59]), in order to reduce the study of real representations to that of complex ones.
Throughout the whole section $\mathfrak{g}$ is a real Lie algebra.

B.1 Preliminaries

First of all we recall the Schur-lemma.

B.1 Proposition (Schur-lemma). *Let κ_1, κ_2 be irreducible representations of $\mathfrak{g}$ on $\mathbb{K}$-vector spaces V_1 and V_2. Let $f \in Hom_{\mathfrak{g}}(V_1, V_2)$ be an invariant homomorphism, i.e.*

$$f \circ \kappa_1(A) = \kappa_2(A) \circ f \qquad \text{for all } A \in \mathfrak{g}.$$

Then holds

1. *f is zero or an isomorphism, i.e. $V_1 \not\simeq V_2$ implies $Hom_{\mathfrak{g}}(V_1, V_2) = 0$.*

2. *If $V_1 = V_2 =: V$ and if f has an eigenvalue $\lambda \in \mathbb{K}$, then $f = \lambda\, id_V$. I.e. if $\mathbb{K} = \mathbb{C}$ and $V_1 = V_2$ we have always $f = \lambda\, id$ with $\lambda \in \mathbb{C}$.*

For invariant bilinear forms, i.e. forms β which satisfy

$$\beta(\kappa(A)u,v)+\beta(u,\kappa(A)v)=0 \qquad \text{for all } A\in\mathfrak{g} \tag{B.1}$$

this gives the following consequence.

B.2 Corollary. *Let κ be an irreducible representation of $\mathfrak{g}$ on a $\mathbb{K}$–vector space V and β be the invariant bilinear form. Then β is zero or non-degenerate.*
If $\mathbb{K}=\mathbb{C}$, then the space of invariant bilinear forms is zero or one-dimensional. It is one-dimensional if and only if $V\simeq_\kappa V^*$. Then it is generated by a symmetric or an anti-symmetric bilinear form.

This consequence is obvious by applying the Schur-lemma to the endomorphism of V, which is induced by two invariant bilinear forms.
For complex representations and invariant sesqui-linear forms, i.e. forms θ with

$$\theta(\lambda u,v)=\lambda\theta(u,v) \qquad \text{and} \qquad \theta(u,\lambda v)=\overline{\lambda}\theta(u,v),$$

one has an analogous result.

B.3 Corollary. *Let κ be an irreducible representation of $\mathfrak{g}$ on a $\mathbb{C}$-vector space V. Every invariant sesqui-linear form is zero or non-degenerate, and the space of invariant sesqui-linear-forms is zero or one-dimensional. It is one dimensional if and only if $\overline{V}\simeq_\kappa V^*$. In this case it is generated by a hermitian or an anti-hermitian form, and the spaces of invariant hermitian and invariant anti-hermitian forms are one-dimensional real subspaces, identified by the multiplication with i.*

In these corollaries we refer to the dual and the conjugate representations, which are defined as follows:

$$\begin{aligned}(\kappa^*(A)\alpha)v &= -\alpha(\kappa(A)v)\\ \overline{\kappa}(A)\overline{v} &= \overline{\kappa(A)v}.\end{aligned}$$

B.4 Definition. Let κ be an arbitrary representation of a Lie algebra $\mathfrak{g}$ on a $\mathbb{K}$-vector space V.

1. Then κ is called **self-dual** if there is an invariant isomorphism between V and V^*. This is equivalent to the existence of an invariant bilinear form β.

2. If $\mathbb{K}=\mathbb{C}$, then κ is called **self-conjugate** if there is an invariant isomorphism from V to $\overline{V}$, i.e. there exists an anti-linear bijective mapping $J: V\longrightarrow V$ which is invariant, i.e. $J\circ\kappa(A)=\kappa(A)\circ J$ for all $A\in\mathfrak{g}$.

It is evident that the existence of an invariant hermitian form or a self-conjugate representation is only possible for **real Lie algebras**.

B.2 Irreducible complex representations of real Lie algebras

B.5 Definition. Let κ be an irreducible complex representation of a real Lie algebra $\mathfrak{g}$ on V. κ is called

of real type if κ is self-conjugate with $J^2 = 1$,

of quaternionic type if κ is self-conjugate with $J^2 = -1$ and

of complex type if κ is not self-conjugate.

From the Schur-lemma it is clear that every complex irreducible representation is either real, complex or quaternionic: If κ is self-conjugate, then J^2 is a linear automorphism of κ so that $J^2 = \lambda\, id$. Furthermore λ must be real because of

$$\lambda Jv = J^2 Jv = JJ^2 v = J\lambda v = \overline{\lambda} Jv.$$

Dividing J by $\sqrt{|\lambda|}$ one gets $\lambda = \pm 1$.
Now it holds

B.6 Proposition. *Let κ be a complex irreducible representation of a real Lie algebra $\mathfrak{g}$. Then κ is of real type if and only if the reellification $\kappa_{\mathbb{R}}$ is reducible.*

Proof. ($\Longrightarrow$) Let κ be of real type, i.e. there is an anti-linear, invariant automorphism of the complex representation space V with J^2 =id. Then J is $\mathbb{R}$-linear, and $V_{\mathbb{R}}$ splits into invariant, real vector spaces

$$\begin{aligned} V_{\pm} &= \{v \in V \mid Jv = \pm v\} \\ V_{\mathbb{R}} &= V_+ \oplus V_-. \end{aligned}$$

So $\kappa_{\mathbb{R}}$ is reducible.
($\Longleftarrow$) Let W be a real, $\kappa_{\mathbb{R}}$-invariant subspace of $V_{\mathbb{R}}$. On $V_{\mathbb{R}}$ the multiplication with i gives an $\mathbb{R}$-automorphism, which defines two subspaces of $V_{\mathbb{R}}$: $W \cap iW$ and $W + iW$. Then both are complex vector spaces in an obvious way, such that they are complex subspaces of V. Since W is $\kappa_{\mathbb{R}}$ invariant, both are κ invariant. Since κ is irreducible, it remains the case that $W \cap iW = \{0\}$ and $W \oplus \mathrm{i}W = V$. But this means that $V = W^{\mathbb{C}}$ such that W defines a conjugation J in V with the desired properties. □

Orthogonal and unitary representations

B.7 Proposition. *Let κ be an irreducible representation of $\mathfrak{g}$. If κ is of complex type, then it cannot be both, unitary and self-dual. If κ is not of complex type, then it is unitary if and only if it is self-dual. In particular one has for real and quaternionic representations (J denotes the automorphism):*

1. *If κ is of real type, then it is orthogonal if and only if it is unitary with respect to θ for which holds $J^*\theta = \overline{\theta}$. It is symplectic if and only if it is unitary with respect to θ satisfying $J^*\theta = -\overline{\theta}$.*

2. *If κ is of quaternionic type, then it is orthogonal if and only if it is unitary with respect to θ with $J^*\theta = -\overline{\theta}$. It is symplectic if and only if it is unitary with respect to θ satisfying $J^*\theta = \overline{\theta}$.*

Proof. Unitary is equivalent to $V^* \simeq_\kappa \overline{V}$ and therefore self-dual is the same as $V \simeq_\kappa \overline{V}$. This gives the proposition. For the remaining single points we get:
1.) Let κ be of real type with respect to a real structure J. By this J one gets from an invariant bilinear form β an invariant sesqui-linear form $\beta(.,J.)$ which is the complex multiple of an invariant hermitian form θ and vice versa. Then one gets for β symmetric/anti-symmetric:

$$\begin{aligned} J^*\theta(u,v) &= \theta(Ju,Jv) = \lambda\beta(Ju,J^2v) \overset{J^2=id}{=} \lambda\beta(Ju,v) \\ &= \pm\lambda\beta(vJu) = \pm\theta(v,u) = \pm\overline{\theta(u,v)}. \end{aligned}$$

2.) analogous with $J^2 = -id$. □

B.8 Corollary. *If κ is positive definite unitary, then it is*

1. *of real type if and only if it is orthogonal,*

2. *of complex type if and only it is not self-dual,*

3. *of quaternionic type if and only if it is symplectic.*

Proof. If θ is positive definite it cannot be $J^*\theta = -\theta$. □

B.3 Irreducible real representations

For a real irreducible representation ρ of a real Lie algebra $\mathfrak{g}$ on a real vector space E two cases are possible: $\rho^{\mathbb{C}}$ is irreducible or reducible. We will describe these cases due to results of E. Cartan ([Car14], see also [Got78], pp.363 and [Iwa59]), in order to reduce the study of real representations to that of complex ones.

B.3.1 Representations of real type

B.9 Proposition. *Let $\mathfrak{g}$ be a real Lie algebra and ρ a representation of $\mathfrak{g}$ on a real vector space E such that $\rho^{\mathbb{C}}$ is* ***irreducible*** *on $E^{\mathbb{C}}$. Then the complex representation $\rho^{\mathbb{C}}$ is of real type.*
If otherwise κ is a complex representation of $\mathfrak{g}$ of ***real type*** *on V, then κ is the complexification of a real irreducible representation of $\mathfrak{g}$.*

Proof. 1.) We show the existence of a $\rho^{\mathbb{C}}$-invariant anti-linear isomorphism J with $J^2 = id$. If we denote by J the conjugation in $E^{\mathbb{C}}$ with respect to E, then it is $J^2 = 1$ and we have

$$J\left(\rho^{\mathbb{C}}(A)(u+iv)\right) = \rho^{\mathbb{C}}(A)(u) - i\rho^{\mathbb{C}}(A)(v) = \rho^{\mathbb{C}}(A)\left(J(u+iv)\right)$$

i.e. J is $\rho^{\mathbb{C}}$-invariant.
2.) In the proof of proposition B.6 we had already shown that for complex representations of real type holds that $V = W^{\mathbb{C}}$. □

So the following definition makes sense.

B.10 Definition. Irreducible real representations with irreducible complexification and irreducible complex representations with reducible reellification (i.e. of real type) are called representations of **real type**.

We have the following correspondence:

$$\begin{array}{rcl} \{\text{real representation of real type}\}_{/\sim} & \leftrightarrow & \{\text{complex representations of real type}\}_{/\sim} \quad (B.2) \\ \rho & \mapsto & \rho^{\mathbb{C}} \\ (\kappa_{\mathbb{R}})_{|\text{maximal invariant subspace}} & \leftarrow & \kappa. \end{array}$$

Here $\sim$ denotes the equivalence of representations.

B.3.2 Representations of non-real type

The situation in this case is described by the following

B.11 Proposition. *Let $\mathfrak{g}$ be a real Lie algebra and ρ be an irreducible representation of $\mathfrak{g}$ on a real vector space E such that $\rho^{\mathbb{C}}$ is* ***reducible*** *on $E^{\mathbb{C}}$.*

1. *If $V \subset E^{\mathbb{C}}$ is any invariant subspace of $\rho^{\mathbb{C}}$. Then holds*

$$E^{\mathbb{C}} = V \oplus \overline{V},$$

where $^{-}$ is the conjugation in $E^{\mathbb{C}}$ with respect to E. V and $\overline{V}$ are irreducible and unique as maximal invariant proper subspaces. The representations on V and $\overline{V}$ are conjugate to each other.

2. *The irreducible representations of $\mathfrak{g}$ on V and on $\overline{V}$ are of complex or of quaternionic type, its reellifications are equivalent to ρ.*

If otherwise κ is a complex irreducible representation of complex or quaternionic type, then κ is the restriction on the maximal invariant proper subspace of the complexification of $\kappa_{\mathbb{R}}$.
If κ is of complex type, then exists and $\kappa_{\mathbb{R}}$–invariant complex structure J on $V_{\mathbb{R}}$. κ is of quaternionic type if and only if there exists and $\kappa_{\mathbb{R}}$–invariant quaternionic structure (I, J, K) on $V_{\mathbb{R}}$.

Proof. 1.) Let $V \subset E^{\mathbb{C}}$ be any invariant, proper subspace of $\rho^{\mathbb{C}}$. Lets denote by $^{-}$ the conjugation in $E^{\mathbb{C}}$ with respect to E.
We consider $W := V + \overline{V}$. Now it is $\overline{W} = W$ which is equivalent to $W = F^{\mathbb{C}}$, where $F = W \cap E$ is a real subspace of E. Since W is invariant under $\rho^{\mathbb{C}}$, F is invariant under ρ. Now ρ is irreducible and therefore $F = E$, i.e. $V + \overline{V} = E^{\mathbb{C}}$. Analogously one shows that $V \cap \overline{V} = \{0\}$, so that one gets

$$V \oplus \overline{V} = E^{\mathbb{C}}.$$

It remains to show that V is irreducible: This is clear since every invariant subspace $U \subset V$ is invariant in $E^{\mathbb{C}}$, but then holds that $U \oplus \overline{U} = E^{\mathbb{C}}$ which implies $U = V$. $\overline{V}$ is irreducible too.
Hence we have two irreducible representations of $\mathfrak{g}$, one on V and one on $\overline{V}$, which are conjugate to each other:

$$\rho^{\mathbb{C}}(A)\overline{v} = \overline{\rho^{\mathbb{C}} v}.$$

So we will denote it by κ and $\overline{\kappa}$.
2.) In order to show that κ and $\overline{\kappa}$ are of complex or quaternionic type, we verify that $\kappa_{\mathbb{R}}$ and $\overline{\kappa}_{\mathbb{R}}$ are irreducible.
For this we show that $\kappa_{\mathbb{R}}$ and $\overline{\kappa}_{\mathbb{R}}$ are isomorphic to ρ. The isomorphism between V and E is given by

$$\begin{array}{rccc} \psi & : & V_{\mathbb{R}} & \longrightarrow & E \\ & & v & \longmapsto & \frac{1}{2}(v + \overline{v}). \end{array}$$

This is obviously an isomorphism of real vector spaces. (Of course this is also an isomorphism between $\overline{V}_{\mathbb{R}}$ and E.) It is also invariant since

$$\psi \circ \kappa_{\mathbb{R}}(A)(x + iy) = \psi(\rho(A)x + i\rho(A)y) = \rho(A)x = \rho(A)(\psi(x + iy)$$

for all $x + iy \in V_{\mathbb{R}}$.
The existence of the complex and the quaternionic structure on $V_{\mathbb{R}}$ is clear. □

Again on defines:

B.12 Definition. Irreducible real representations with reducible complexification and irreducible complex representations with irreducible reellification are called representations of **non-real type** (of **complex** or **quaternionic type** respectively).

Again we have the correspondence

$$\left\{ \begin{array}{l} \text{real representations} \\ \text{of non-real type} \end{array} \right\}_{/\sim} \leftrightarrow \left\{ \begin{array}{l} \text{complex representations} \\ \text{of non-real type} \end{array} \right\}_{/\approx} \tag{B.3}$$

$$\begin{array}{ccl} \rho & \mapsto & \rho^{\mathbb{C}}_{|\text{maximal invariant subspace}} \\ \kappa_{\mathbb{R}} & \leftarrow\!\!\shortmid & \kappa. \end{array}$$

Here $\sim$ denotes the equivalence of representation and $\approx$ the equivalence

$$\kappa_1 \approx \kappa_2 \quad \Leftrightarrow \quad \kappa_1 \sim \kappa_2 \text{ or } \kappa_1 \sim \overline{\kappa_2}.$$

On the real space $E \simeq V_{\mathbb{R}}$ we have the complex structure J, i.e. an $\mathbb{R}$–automorphism with $J^2 = -1$ given by the multiplication with i: $Jv = iv$. J commutes with ρ since

$$\rho(A)(Jv) = \kappa_{\mathbb{R}}(A)(Jv) = \kappa(A)iv = i\kappa(A)v = J(\kappa(A)v).$$

One describes the complex vector space V as a subspace in $E^{\mathbb{C}}$ as follows. One extends the complex structure to an automorphism of $E^{\mathbb{C}}$ also denoted by J and with the property $J^2 = -1$. Then one defines

$$V_{\pm} := \{v \in E^{\mathbb{C}} | Jv = \pm i\, v\} \subset E^{\mathbb{C}}$$

and gets $E^{\mathbb{C}} = V_+ \oplus V_-$. Furthermore it is

$$V_{\pm} = \{x \mp iJx | x \in E\} \text{ and therefore } V_{\pm} = \overline{V_{\mp}}. \tag{B.4}$$

Then one has the following isomorphisms, invariant under the corresponding representations:

$$\begin{array}{ccccccc} E & \simeq_{\mathbb{R}} & V & \simeq_{\mathbb{C}} & V_+ & \simeq_{\mathbb{R}} & V_- = \overline{V_+} \\ \frac{1}{2}(v + \overline{v}) & \leftarrow\!\!\shortmid & v & \mapsto & \frac{1}{2}(v - iJv) & \mapsto & \frac{1}{2}(v + iJv). \end{array} \tag{B.5}$$

B.3.3 Weights of real type and non-real type representations

For sake of completeness we will here cite a result of Iwahori describing the real type and non-real type property in terms of weights.

If κ is a complex representation of a real Lie algebra $\mathfrak{g}$ then one can extend κ linearly to $\mathfrak{g}^{\mathbb{C}}$ and it holds that κ is irreducible if and only if its extension to $\mathfrak{g}^{\mathbb{C}}$ is irreducible.

The conjugate representation Let κ be a complex representation of a real Lie algebra $\mathfrak{g}$.

Let $\mathfrak{g}^{\mathbb{C}}$ be the complexification of $\mathfrak{g}$, and A the real structure of $\mathfrak{g}^{\mathbb{C}}$ with respect to the real form $\mathfrak{g}$. Let $\mathfrak{h}$ be the Cartan subalgebra of $\mathfrak{g}^{\mathbb{C}}$. To every $\lambda \in \mathfrak{h}^*$ one defines $\overline{\lambda} \in \mathfrak{h}$ by

$$\overline{\lambda}(H) := \overline{\lambda(A(H))}. \tag{B.6}$$

Let Δ be tho root system of $\mathfrak{g}^{\mathbb{C}}$. If α is a root then $\overline{\alpha}$ too and it is $\langle \alpha, \beta \rangle = \langle \overline{\alpha}, \overline{\beta} \rangle$. If Π is a fundamental root system, then $\overline{\Pi}$ too. This means that there is an element σ' of the Weyl group, such that $\sigma'(\Pi) = \overline{\Pi}$.
If now the extension of the representation κ of $\mathfrak{g}$ to $\mathfrak{g}^{\mathbb{C}}$ is of highest weight Λ then the extension of the conjugate representation is of highest weight $\sigma'(\overline{\Lambda})$.

Hermitian representations: In terms of the highest weights one gets the following

B.13 Proposition. *Let κ be a irreducible representation of a real Lie algebra $\mathfrak{g}$ of highest weight Λ. Let σ resp. σ' be the elements of the Weyl group such that $\sigma(\Delta_+) = \Delta_-$ resp. $\sigma'(\Pi) = \overline{\Pi}$. Then κ is hermitian if and only if*

$$\sigma(\Lambda) = -\sigma'(\overline{\Lambda}).$$

Self conjugate representations: In terms of highest weights self conjugateness is equivalent to $\Lambda = \sigma'(\overline{\Lambda})$.

Let κ be irreducible, complex and $\Lambda = \sum_{i=1}^{n} m_i \omega_i$ be the highest weight of κ with ω_i the fundamental weights. Let κ_{ω_i} be the corresponding fundamental representations. These should be ordered that

$$\kappa_{\omega_1} \sim \overline{\kappa}_{\omega_2}, \dots, \kappa_{\omega_{2k-1}} \sim \overline{\kappa}_{\omega_{2k}}, \kappa_{\omega_{2k+1}} \sim \overline{\kappa}_{\omega_{2k+1}}, \dots, \kappa_{\omega_n} \sim \overline{\kappa}_{\omega_n}.$$

From the above fact that κ is self conjugate if $\Lambda = \sigma'(\overline{\Lambda})$ one gets that κ is self conjugate if and only if

$$m_1 = m_2, \dots, m_{2k-1} = m_{2k}.$$

Let now $\varepsilon_{2k+1}, \dots, \varepsilon_n$ be the indices of the self conjugate under the fundamental representation. Then if κ is self conjugate then its index is $\varepsilon = \varepsilon_{2k+1}^{m_{2k+1}} \cdot \ldots \cdot \varepsilon_n^{m_n}$.
Hence one has the following

B.14 Theorem. *[Iwa59] Let ρ be a real representation of a real Lie algebra $\mathfrak{g}$, and $\kappa := \rho^{\mathbb{C}}$ of highest weight $\Lambda = \sum_{i=1}^{n} m_i \omega_i$, the m_i ordered as above. Then ρ is of real type if and only if*

$$m_1 = m_2, \dots, m_{2k-1} = m_{2k}$$

and $\varepsilon_{2k+1}^{m_{2k+1}} \cdot \ldots \cdot \varepsilon_n^{m_n} = 1.$

B.4 Orthogonal real representations

Let now ρ be a real representation of $\mathfrak{g}$ on E which should be orthogonal (or symplectic) with respect to a (anti-)symmetric bilinear form h.

On $E^{\mathbb{C}}$ h defines a (anti-)symmetric bilinear form by bilinear extension, denoted by $h^{\mathbb{C}}$ and a (anti-)hermitian form by conjugate linear extension in the second component, denoted by h'. Both are invariant under $\rho^{\mathbb{C}}(\mathfrak{g})$. The hermitian form has the same signature as the symmetric form h. The existence of an invariant anti-hermitian form is equivalent to the existence of an invariant hermitian form.
For the conjugation in $E^{\mathbb{C}}$ we have the following relations

$$h'(u,v) = h^{\mathbb{C}}(u,\bar{v}) = \overline{h^{\mathbb{C}}(\overline{u},v)} = \overline{h'(v,u)} \ .$$

B.4.1 Orthogonal and symplectic representations of real type

From these introductory remarks we obtain the following proposition for real type representations which can be found in [Ber55](for the orthogonal case), and is standard now.

B.15 Proposition. *[Ber55] Let ρ be a real representation of real type of a real Lie Algebra $\mathfrak{g}$ on a real vector space E, orthogonal or symplectic with respect to h. Let β^h denote the complex linear and θ^h the hermitian extension of h on $V = E^{\mathbb{C}}$. Then both are non-degenerate and $\rho^{\mathbb{C}}$ is orthogonal/symplectic with respect to β^h and unitary with respect to θ^h. θ^h has the same index as h in case h is orthogonal.*

This gives a

B.16 Corollary. *If ρ is of real type, then the space of invariant bilinear form is one-dimensional and generated by a symmetric or an anti-symmetric form.*

The *proof* is clear because the irreducibility of $\rho^{\mathbb{C}}$ gives that $h_1^{\mathbb{C}} = h_2^{\mathbb{C}}$, which implies $h_1 = h_2$. □

We will now prove the other direction of proposition B.15.

B.17 Proposition. *Let $\mathfrak{g}$ be a real Lie algebra and κ an irreducible, complex representation of real type on V, which decomposes $\kappa_{\mathbb{R}}$-invariant into $V = E \oplus iE$, and set $\rho = (\kappa_{\mathbb{R}})_{|E}$ the corresponding irreducible real representation. If κ is unitary (and therefore self-dual), then ρ is self-dual, i.e. orthogonal or symplectic and we have two cases:*

1. *If κ is orthogonal, then ρ is orthogonal.*
2. *If κ is symplectic, then ρ is symplectic*

Proof. Let κ be unitary with respect to θ, which defines two bilinear mappings on E

$$\begin{aligned} h_1(x,y) &= Re\,(\theta(x,y)) \text{ symmetric} \\ h_2(x,y) &= Im\,(\theta(x,y)) \text{ anti-symmetric.} \end{aligned}$$

Both are ρ invariant. If both are degenerate, then both are zero by the Schur-lemma and so θ must be zero, which is a contradiction.
1.) If in addition κ is orthogonal, then for θ holds by proposition B.7 that $J^*\theta = \overline{\theta}$, where J is the conjugation of E in $E^{\mathbb{C}}$. But in this case h_2 is zero, because $E = \{v \in V | Jv = v\}$:

$$h_2(x,y) = Im\theta(x,y) = Im\theta(Jx,Jy) = Im\overline{\theta(x,y)} = -Im\theta(x,y) = -h_2(x,y).$$

Hence h_1 must be non degenerate and therefore ρ orthogonal.
2.) If κ is symplectic one shows analogously with proposition B.7 that $h_1 = 0$ and therefore ρ symplectic. □

Both results give the following equivalence:

$$\{\rho \text{ real, real type, self-dual}\}_{/\sim} \quad \leftrightarrow \quad \left\{ \begin{array}{l} \kappa \text{ complex, real type,} \\ \text{self-dual} \mathrel{\hat{=}} \text{unitary} \end{array} \right\}_{/\sim} \tag{B.7}$$

$$\left\{ \begin{array}{l} \rho \text{ real, real type,} \\ \text{orthogonal/symplectic} \end{array} \right\}_{/\sim} \quad \leftrightarrow \quad \left\{ \begin{array}{l} \kappa \text{ complex, real type,} \\ \text{orthogonal/symplectic} \end{array} \right\}_{/\sim}. \tag{B.8}$$

B.4.2 Orthogonal representations of non-real type

For non-real type representations we have the $\rho^{\mathbb{C}}$-invariant decomposition $E^{\mathbb{C}} = V \oplus \overline{V}$. In a basis, adapted to this decomposition $h^{\mathbb{C}}$ and h' are given as follows

$$h^{\mathbb{C}} = \begin{pmatrix} A & B \\ B^t & \overline{A} \end{pmatrix} \quad \text{and} \quad h' = \begin{pmatrix} B & A \\ \overline{A} & B^t \end{pmatrix}$$

where $A = A^t$ and $B^t = \overline{B}$ are quadratic matrices with the dimension of V.
Now one defines a bilinear and a sesqui-linear form on V resp. on $\overline{V}$:

$$\begin{array}{rclcll} \beta^h(u,v) & := & h^{\mathbb{C}}(u,v) & = & h'(u,\overline{v}) & \text{symmetric/anti-symmetric} \\ \theta^h(u,v) & := & h^{\mathbb{C}}(u,\overline{v}) & = & h'(u,v) & \text{hermitian/anti-hermitian} \end{array}$$

for $u, v \in V$ resp. $\overline{V}$. Both are invariant under $\kappa = \rho^{\mathbb{C}}_{|V}(\mathfrak{g})$.
From the Schur-lemma it is clear that at least one of them is non-degenerate, since $h^{\mathbb{C}}$ is non-degenerate.
Using the isomorphisms of (B.5) we can give θ^h and β^h explicitly:

$$\beta^h(x - iJx, y - iJy) = \frac{1}{4}\left(h(x,y) - h(Jx,Jy) - i\left(h(Jx,y) + h(x,Jy)\right)\right) \tag{B.9}$$

$$\theta^h(x - iJx, y - iJy) = \frac{1}{4}\left(h(x,y) + h(Jx,Jy) + i\left(h(x,Jy) - h(Jx,y)\right)\right) \tag{B.10}$$

Again we have the proposition of Berger (for the orthogonal case).

B.18 Proposition. *[Ber55] Let ρ be a real orthogonal/symplectic representation of non-real type, i.e. $(E,\rho) = (V_\mathbb{R}, \kappa_\mathbb{R})$. Then κ is invariant under β^h and θ^h and at least one of them is non-degenerate, i.e. κ is orthogonal/symplectic or unitary/anti-unitary with respect to β^h or θ^h.*
Furthermore holds: If $\mathfrak{g}$ contains a real sub-algebra $\mathfrak{h} \neq 0$ such that $\mathfrak{h} = \mathfrak{p}_\mathbb{R}$ where $\mathfrak{p}$ is a complex Lie algebra, then $\theta^h = 0$ i.e. β^h non-degenerate.

Proof. We only have to prove the second assertion.
By assumption we have a complex Lie structure on $\mathfrak{h}$, i.e. a automorphism J with $J^2 = -1$ and $J \circ ad_X = ad_X \circ J$. As above for vector spaces we have here a Lie algebra decomposition

$$\mathfrak{g}^\mathbb{C} \supset \mathfrak{h}^\mathbb{C} = \mathfrak{p}_+ \oplus \mathfrak{p}_- \text{ with } \mathfrak{p}_\pm = \{v \in \mathfrak{h}^\mathbb{C} | Jv = \pm iv\}.$$

Then $\mathfrak{p} \simeq_\mathbb{C} \mathfrak{p}_+$.
Let now $\rho^\mathbb{C}$ be extended to $\mathfrak{g}^\mathbb{C}$. Then because of its linearity $h^\mathbb{C}$ is invariant under $\rho^\mathbb{C}(\mathfrak{g}^\mathbb{C})$. But if we suppose that θ^h is invariant under $\mathfrak{g}$ we have for a $H \in \mathfrak{h}$ and $\kappa = \rho^\mathbb{C}_{|V}$ as above

$$\begin{aligned}
0 &= \theta^h(\kappa(JH)v, w) + \theta^h(v, \kappa(JH)w) \\
&\overset{p.d.}{=} h^\mathbb{C}(\kappa(JH)v, \overline{w}) + h^\mathbb{C}(v, \overline{\kappa(JH)w}) \\
&= h^\mathbb{C}(\rho^\mathbb{C}(JH)v, \overline{w}) + h^\mathbb{C}(v, \overline{\rho^\mathbb{C}(JH)w}) \\
&\overset{H\in\mathfrak{p}_+}{=} i\left(h^\mathbb{C}(\rho^\mathbb{C}(H)v, \overline{w}) - h^\mathbb{C}(v, \overline{\rho^\mathbb{C}(H)w})\right) \\
&= i\left(\theta^h(\kappa(H)v, w) - \theta^h(v, \kappa(H)w)\right) \\
&\overset{\theta^h \text{ invariant}}{=} 2i\theta^h(\kappa(H)v, w)
\end{aligned}$$

for all $H \in \mathfrak{h}$, $v, w \in V$. This means $\mathfrak{h} \subset ker\ \kappa = 0$. □

We also can show the other direction.

B.19 Proposition. *Let $\mathfrak{g}$ be a real Lie algebra, κ be a complex representation of non-real type (of complex or quaternionic type), i.e. $\rho = \kappa_\mathbb{R}$ is irreducible. Then holds:*

1. *If κ is unitary with respect to θ or orthogonal with respect to β, then ρ is orthogonal with respect to h and $\theta^h = \theta$ or $\beta^h = \beta$.*
2. *If κ is anti-unitary with respect to θ or symplectic with respect to β, then ρ is symplectic with respect to h and $\theta^h = \theta$ or $\beta^h = \beta$.*

Proof. We define a bilinear form on $E = V_\mathbb{R}$ by

$$h(x,y) := Re\ \theta(x - iJx, y - iJy) \quad \text{or} \quad h(x,y) := Re\ \beta(x - iJx, y - iJy).$$

This form is invariant and — since $Re\ iz = -Im\ z$ — also non-degenerate. (The difference to real type is that here the arguments in θ/β run over the whole complex vector space V.) h is symmetric if κ is unitary or orthogonal and anti-symmetric if β is anti-symmetric or anti-unitary. The fact that the extensions are equal to θ resp. β follows from the formulas (B.10) and (B.9). □

Again we have the following correspondence:

$$\{\rho \text{ real, non-real type, orthogonal}\}_{/\sim} \quad \leftrightarrow \quad \left\{ \begin{array}{l} \kappa \text{ complex, non-real type,} \\ \text{unitary or orthogonal} \end{array} \right\}_{/\approx} \tag{B.11}$$

$$\{\rho \text{ real, non-real type, symplectic}\}_{/\sim} \quad \leftrightarrow \quad \left\{ \begin{array}{l} \kappa \text{ complex, non-real type,} \\ \text{symplectic or anti-unitary} \end{array} \right\}_{/\approx} \tag{B.12}$$

The fact that a complex representation is unitary if and only if it is anti-unitary (the anti-hermitian form is $i\theta$) implies that a real, orthogonal representation of non-real type with non-degenerate θ^h on the corresponding complex representation is also symplectic. This corresponds to the equality of real matrix algebras:

$$\begin{aligned} \mathfrak{u}(n) &= \mathfrak{so}(2n) \cap \mathfrak{sp}(2n) \\ &= \{X \in \mathfrak{gl}(2n)\,|X^t = -X\} \cap \left\{ \left(\begin{array}{cc} A & B \\ C & -A^t \end{array} \right) \middle| A, B, C \in \mathfrak{gl}(n), B^t = B, C^t = C \right\} \\ &= \left\{ \left(\begin{array}{cc} A & B \\ -B & A \end{array} \right) \middle| A, B \in \mathfrak{gl}(n), A^t = -A, B^t = B, \right\}. \end{aligned}$$

I.e. if a complex representation κ of non-real type is unitary, then $\kappa_{\mathbb{R}}$ is orthogonal and symplectic.
Furthermore one proves the following

B.20 Lemma. *Let h be symmetric, β^h, θ^h as above and J the complex structure on E. Then holds*

1. *$\beta^h = 0$ if and only if $h(x, y) = h(Jx, Jy)$ for all $x, y \in E$.*
2. *$\theta^h = 0$ if and only if $h(x, y) = -h(Jx, Jy)$ for all $x, y \in E$.*

Proof. If we write every element of $V = V_+$ in the form (B.4) we get the proposition due to formulas (B.9) and (B.10). □

We will now prove the main result for the case that h is positive definite.

B.21 Proposition. *Let ρ be irreducible of non-real type and orthogonal with respect to h where h is* ***positive definite****.*

Then the corresponding complex representation κ of non-real type is unitary, with respect to a positive definite hermitian form, which is the standard hermitian form for representations of compact Lie groups/Lie algebras.
κ is not orthogonal, i.e. the linear extension β^h of h vanishes on $V \times V$.

Proof. We can prove this in two ways.
If θ^h is degenerate, then it is zero and we have by lemma B.20 that $h(x,x) = -h(JxJx)$. But this is not possible if h is positive definite. So θ^h is non degenerate and by formula (B.10) positive definite, since h is positive definite. But the existence of a positive definite hermitian form entails by corollary B.8 for non-real type representations, i.e. of complex or quaternionic type, that the representation cannot be orthogonal. So $\beta^h = 0$.
An easier way to argue is that representations of compact Lie algebras are unitary with respect to a standard positive definite hermitian form. This form is unique and thats why equal to θ^h and by corollary B.8 the representation cannot be orthogonal. □

Appendix C

Algebras with first prolongation V^* (Proof of proposition 4.59)

For the Lie algebras of table 2 on page 137 we will now verify that they are weak-Berger, but also Berger algebras, and that they are the holonomy of a Riemannian symmetric space which is Kähler. We will check the following points

(i) We will give the isomorphism between $\overline{V}$ and $\mathfrak{g}^{(1)}$. We will not prove that it is an isomorphism, here we refer to the result of [KN65]. To see that the homomorphism we will give is injective is easy in all the cases, to see that it is surjective is much more difficult.

(ii) With the help of this isomorphism we will show that $\mathfrak{g} \subset \tilde{\mathfrak{g}}$, i.e. that the compact real form — which is in all cases unique, because the Lie algebras $\mathfrak{g}$ are all reductive — is weak-Berger.

(iii) Further we will show that, with respect to conjugation defined by this compact real form, the isomorphism satisfies the condition (4.60) from the corollary 4.58, which implies that they are Berger algebras too.

(iv) Finally we will give in each case a Riemannian symmetric space with holonomy $\mathfrak{g}_0$.

C.1 $\mathfrak{co}(n, \mathbb{C})$ acting on $V := \mathbb{C}^n$

If we denote by $\langle .,. \rangle$ a symmetric, bilinear form on V, we can define its conformal algebra $\mathfrak{co}(V, \langle .,. \rangle) := \mathbb{C}Id \oplus \mathfrak{so}(V, \langle .,. \rangle) = \{A \in \mathfrak{gl}(V) | \langle Au, v\rangle + \langle u, Av\rangle = \lambda\langle u, v\rangle\}$. If we fix an orthonormal basis $(e_1, \dots, e_n)$ we can write it in matrices

$$\mathfrak{co}(V, \langle .,. \rangle) = \mathfrak{co}(n, \mathbb{C}) = \mathbb{C}E_n \oplus \mathfrak{so}(n, \mathbb{C})$$

with E_n the unit matrix.
The compact real form of this algebra is of course

$$\mathfrak{g}_0 \;:=\; i\mathbb{R}E_n \oplus \mathfrak{so}(n,\mathbb{R}) \simeq \mathfrak{so}(2,\mathbb{R}) \oplus \mathfrak{so}(n,\mathbb{R}).$$

The reellification of this representation is generated by the matrices

$$J := \begin{pmatrix} 0 & I_n \\ -I_n & 0 \end{pmatrix} \quad \text{and} \quad \begin{pmatrix} A & 0 \\ 0 & A \end{pmatrix} \quad \text{with } A \in \mathfrak{so}(n,\mathbb{R})$$

in $\mathfrak{so}(2n,\mathbb{R})$. This representation is irreducible, i.e. $\mathfrak{g}_0$ is of type two.
So we have the $\mathfrak{g}_0$-invariant hermitian form θ on $\mathbb{C}^n$ such that $\mathfrak{g}_0 \subset \mathfrak{u}(n) \subset \mathfrak{gl}(n,\mathbb{C})$. In particular we have $\mathfrak{g}_0 = \mathfrak{u}(n) \cap \mathfrak{co}(n,\mathbb{C})$. So we have that the orthonormal basis $(e_1,\ldots,e_n)$ is also an orthonormal basis for θ.
We will now check the points.

(i) The isomorphism between $\overline{V} \overset{\beta}{\simeq} \mathfrak{co}(n,\mathbb{C})^{(1)}$ is defined by

$$\mathfrak{co}(n,\mathbb{C}) \ni \beta(\overline{u},v) := Q_{\theta(.,u)}(v) \;:\; w \longmapsto \theta(v,u)w + \theta(w,u)v - \langle v,w\rangle u^*,$$

where u^* denotes the element in V, defined by the relation

$$\langle u^*, w\rangle \;=\; \theta(w,u) \qquad\qquad \text{for all } w \in V.$$

Clearly * is a conjugate linear, bijective mapping on V.
Then it is $\beta(\overline{u},v) \in \mathfrak{co}(n,\mathbb{C})$ since

$$\begin{aligned}
&\langle \beta(\overline{u},v)w, w'\rangle + \langle w, \beta(\overline{u},v)w'\rangle = \\
&= \theta(v,u)\langle w,w'\rangle + \langle v,w'\rangle \underbrace{(\theta(w,u) - \langle u^*,w\rangle)}_{=0} + \langle v,w\rangle \underbrace{(\theta(w',u) - \langle u^*,w'\rangle)}_{=0} \\
&= 2\theta(v,u)\langle w,w'\rangle.
\end{aligned}$$

$Q_{\theta(.,u)}$ is also in $\mathfrak{g}^{(1)}$ because of the symmetry in v and w.

(ii) Weak-Berger: We show that the elements $Q_{\theta(.,u)}(v)$ generate the whole $\mathfrak{co}(n,\mathbb{C})$. Therefore we consider the $\langle .,.\rangle$-orthonormal base $e_1,\ldots,e_n$. Since this is also an orthonormal basis for θ, and we have that $e_i^* = e_i$. This fact gives that

$$\begin{aligned}
Q_{\theta(.,e_i)}(e_j) \;:\; w &\mapsto \theta(e_i,e_i)w + \theta(w,e_i)e_j - \theta(w,e_j)e_i \quad \text{i.e.} \\
Q_{\theta(.,e_i)}(e_i) &= Id \\
Q_{\theta(.,e_i)}(e_j) &= E_{ij} \;:\; e_k \mapsto \delta_{ki}e_j - \delta_{kj}e_i \;\text{ for } i \neq j,
\end{aligned}$$

where E_{ij} is the standard basis of $\mathfrak{so}(n,\mathbb{C})$ with respect to $e_1,\ldots,e_n$. So we have $\mathfrak{co}(n,\mathbb{C}) \subset \widetilde{\mathfrak{co}(n,\mathbb{C})}$.

(iii) Berger: Here we work again with the basis elements. For the conjugation in $\mathfrak{co}(n,\mathbb{C})$ with respect to $\mathfrak{g}_0 = i\mathbb{R}Id \oplus \mathfrak{so}(n,\mathbb{R})$ one gets that

$$\overline{Id} = -Id \qquad \text{and} \qquad \overline{E_{ij}} = E_{ij}.$$

But with this conjugation β satisfies the relation (4.60):

$$\begin{array}{rclclcl} \overline{Q_{\theta(.,e_i)}(e_i)} & = & \overline{Id} & = & -Id & = & -Q_{\theta(.,e_i)}(e_i) \\ \overline{Q_{\theta(.,e_i)}(e_j)} & = & \overline{E_{ij}} & = & E_{ij} = -E_{ji} & = & -Q_{\theta(.,e_j)}(e_i). \end{array}$$

(iv) The Riemannian symmetric space: We consider the compact, Kählerian, Riemannian symmetric space of type $BD\ I$ given by the quotient $SO(2+n)/SO(2) \times SO(n)$. (For this and the following symmetric spaces see [Hel78].)
Its tangent space is equal to

$$\begin{array}{rcl} \mathfrak{so}(n+2,\mathbb{R})/\mathfrak{so}(2,\mathbb{R}) \oplus \mathfrak{so}(n,\mathbb{R}) & \simeq & \mathbb{R}^{2n} \\ \left[\begin{pmatrix} 0 & a & X \\ -a & 0 & Y \\ -X & -Y & B \end{pmatrix}\right] & \mapsto & (X,Y), \end{array}$$

with $X, Y \in \mathbb{R}^n$, $a \in \mathbb{R}$ and $B \in \mathfrak{so}(n,\mathbb{R})$. Therefore the holonomy representation of $\mathfrak{so}(2,\mathbb{R}) \oplus \mathfrak{so}(n,\mathbb{R}) \simeq \mathfrak{co}(n,\mathbb{R})$ on $\mathbb{R}^{2n}$, given by the adjoint representation is

$$\begin{array}{rcl} \left(\mathfrak{so}(2,\mathbb{R}) \oplus \mathfrak{so}(n,\mathbb{R}), \mathbb{R}^{2n}\right) & \longrightarrow & \mathbb{R}^{2n} \\ \begin{pmatrix} 0 & 1 \\ -1 & 0 \end{pmatrix} \cdot (X,Y) & = & (Y,-X) \\ B \cdot (X,Y) & = & (B \cdot X, B \cdot Y). \end{array}$$

But this representation is the same as the representation of $\mathfrak{co}(n,\mathbb{R})$ on $(\mathbb{C}^n)_\mathbb{R} = \mathbb{R}^{2n}$ which is the reellification of the first entry of table 1.
(Here — as well as in the following — the same can be performed for the non-compact symmetric space $SO_0(2,n)/SO(2) \times SO(n)$ of type $BD\ I$. See also [Hel78])
All together gives that $\mathfrak{g}_0$ is weak-Berger, but Berger too and the holonomy algebra of a Riemannian symmetric space.

C.2 $\mathfrak{gl}(n,\mathbb{C})$ acting on $V := \odot^2\mathbb{C}^n$

The compact real form is $\mathfrak{u}(n)$ and the representations on $\odot^2\mathbb{C}^n$ is of weight $2\omega_1$ and therefore of type 2 (see [Iwa59]). Thats why we have that

$$\mathfrak{u}(n) \subset \mathfrak{u}(\odot^2\mathbb{C}^n, \theta) \subset \mathfrak{gl}(\odot^2\mathbb{C}^n).$$

If θ_0 is the hermitian on $\mathbb{C}^n$, then θ is given via θ_0 obviously by

$$\theta(uv,rs) = \theta_0(u,r)\cdot\theta_0(v,s) + \theta_0(u,s)\cdot\theta_0(v,r).$$

Hence a θ_0-orthonormal base $e_1,\dots,e_n$ in $\mathbb{C}^n$ defines a θ-orthonormal base $(e_ie_j, i\le j)$ in $\odot^2\mathbb{C}^n$.
Let D_{ij} be the standard basis of $\mathfrak{gl}(n)$, mapping $e_k \mapsto \delta_{ik}e_j$. On $\odot^2\mathbb{C}^n$ it acts as follows

$$D_{ij} \;:\; e_pe_q \;\longmapsto\; (\delta_{ip}e_q + \delta_{iq}e_p)\, e_j.$$

Again we check the three points.

(i) The isomorphism between $\overline{V} \overset{\beta}{\simeq} \mathfrak{gl}(n,\mathbb{C})^{(1)}$ is defined by

$$\begin{aligned}\mathfrak{gl}(\odot^2\mathbb{C}^n) \supset \mathfrak{gl}(n,\mathbb{C}) \ni \beta(\overline{\sigma},uv) &:= Q_{\theta(.,\sigma)}(uv) :\\ xy \;\longmapsto\; &\frac{1}{2}\left(\theta(ux,\sigma)vy + \theta(vx,\sigma)uy + \theta(uy,\sigma)vx + \theta(vy,\sigma)ux\right)\end{aligned}$$

for $\sigma, uv, xy \in V$. Then it is $\beta(\overline{\sigma},\nu) \in \mathfrak{gl}(n,\mathbb{C}) \subset \mathfrak{gl}(\odot^2\mathbb{C}^n)$ and $Q_{\theta(.,\sigma)}$ is also in the first prolongation because of the symmetry in the definition.

(ii) Weak-Berger: We show that the elements $Q_{\theta(.,uv)}(rs)$ generate the whole $\mathfrak{gl}(n,\mathbb{C})$. A direct calculation with the above θ-orthonormal base e_ie_j and D_{ij}, then gives

$$Q_{\theta(.,e_ie_j)}(e_ke_l) = \delta_{ik}D_{jl} + \delta_{jk}D_{il} + \delta_{il}D_{jk} + \delta_{jl}D_{ik}. \tag{C.1}$$

So we get that

$$Q_{\theta(.,e_ie_j)}(e_je_j) \;=\; Q_{\theta(.,e_je_j)}(e_ie_j) \;=\; D_{ij} + \delta_{ij}D_{jj},$$

which shows that the whole $\mathfrak{gl}(n,\mathbb{C})$ is generated by $\beta(\overline{u},v)$. So we have $\mathfrak{gl}(n,\mathbb{C}) \subset \widetilde{\mathfrak{gl}(n,\mathbb{C})} \subset \mathfrak{gl}(\odot^2\mathbb{C}^n)$.

(iii) Berger: The standard basis of $\mathfrak{u}(n) \in \mathfrak{gl}(n,\mathbb{C})$ is the following

$$\frac{i}{2}(D_{ij}+D_{ji})\,,\;\; \frac{1}{2}(D_{ij}-D_{ji}) \;\text{ for } k=1,\dots,n, 1\le i\le j\le n.$$

So we have the conjugation with respect to $\mathfrak{u}(n)$:

$$\begin{aligned}\overline{D_{ij}} &= \overline{\frac{1}{2}(D_{ij}-D_{ji}) - i\frac{i}{2}(D_{ij}+D_{ji})}\\ &= \frac{1}{2}(D_{ij}-D_{ji}) + i\frac{i}{2}(D_{ij}+D_{ji}) \;=\; -D_{ji}.\end{aligned}$$

But with this conjugation β satisfies the relation (4.60):

$$\begin{aligned}\overline{Q_{\theta(.,e_ie_j)}(e_ke_l)} &= -\left(\delta_{ik}D_{lj} + \delta_{jk}D_{li} + \delta_{il}D_{kj} + \delta_{jl}D_{ki}\right)\\ &= -Q_{\theta(.,e_ke_l)}(e_ie_j).\end{aligned}$$

(iv) The Riemannian symmetric space: We consider the compact, Kählerian, Riemannian symmetric space of type $C\ I$ given by the quotient $Sp(2n,\mathbb{R})/U(n)$. For tangent space we have

$$\mathfrak{sp}(2n,\mathbb{R})/\mathfrak{u}(n) \simeq \left\{ \left(\begin{array}{cc} R & S \\ S & -R \end{array} \right) \middle| R, S \in \mathfrak{gl}(n,\mathbb{R}) \text{ symmetric} \right\} \simeq \mathbb{R}^{n(n+1)}.$$

The holonomy representation is given by the adjoint representation of $\mathfrak{u}(n)$ on these matrices:

$$\begin{aligned} \underbrace{\left(\begin{array}{cc} A & B \\ -B & A \end{array} \right)}_{\in \mathfrak{u}(n)} \cdot \left(\begin{array}{cc} R & S \\ S & -R \end{array} \right) &= \left[\left(\begin{array}{cc} A & B \\ -B & A \end{array} \right), \left(\begin{array}{cc} R & S \\ S & -R \end{array} \right) \right] \\ &= \left(\begin{array}{cc} [A,R]+BS+SB & [A,S]-BR-RB \\ [A,S]-BR-RB & -[A,R]-BS-SB \end{array} \right). \end{aligned}$$

A direct calculation shows that this representation is isomorphic to the reellification of the representation in the second entry of table 2 via the following isomorphism

$$\begin{aligned} \left(\odot^2\mathbb{C}^n\right)_{\mathbb{R}} &\simeq \left\{ \left(\begin{array}{cc} R & S \\ S & -R \end{array} \right) \middle| R, S \in \mathfrak{gl}(n,\mathbb{R}) \text{ symmetric} \right\} \\ e_i \cdot e_j &\mapsto \left(\begin{array}{cc} B_{ij} & 0 \\ 0 & -B_{ij} \end{array} \right) \\ i\ e_i \cdot e_j &\mapsto \left(\begin{array}{cc} 0 & B_{ij} \\ B_{ij} & 0 \end{array} \right) \end{aligned}$$

where B_{ij} is the standard basis of symmetric matrices.
All together gives that $\mathfrak{g}_0$ is weak-Berger, but Berger too and the holonomy algebra of the Riemannian symmetric space of type $C\ I$.

C.3 $\mathfrak{gl}(n,\mathbb{C})$ acting on $V := \wedge^2\mathbb{C}^n$

The representations on $\wedge^2\mathbb{C}^n$ is of weight ω_2 which is of non-real type for $n \geq 5$. We have again

$$\mathfrak{u}(n) \subset \mathfrak{u}(\wedge^2\mathbb{C}^n, \theta) \subset \mathfrak{gl}(\wedge^2\mathbb{C}^n)$$

and

$$\theta(u \wedge v, r \wedge s) = \theta_0(u,r) \cdot \theta_0(v,s) - \theta_0(u,s) \cdot \theta_0(v,r).$$

Thus a θ_0-orthonormal base $e_1, \ldots, e_n$ in $\mathbb{C}^n$ defines a θ-orthonormal base $(e_i \wedge e_j, i < j)$ in $\wedge^2\mathbb{C}^n$.

The standard basis D_{ij} of $\mathfrak{gl}(n)$ acts as follows

$$D_{ij} \ : \ e_p \wedge e_q \ \longmapsto \ (\delta_{iq} e_p - \delta_{ip} e_q) \wedge e_j.$$

Again we check the three points.

(i) The isomorphism between $\overline{V} \overset{\beta}{\simeq} \mathfrak{gl}(n,\mathbb{C})^{(1)}$ is defined by

$$\begin{aligned} \mathfrak{gl}(\wedge^2\mathbb{C}^n) \supset \mathfrak{gl}(n,\mathbb{C}) \ni \beta(\overline{\omega}, u \wedge v) := Q_{\theta(.,\omega)}(u \wedge v) \ : & \\ x \wedge y \ \longmapsto \ & \frac{1}{2}(\theta(u \wedge x, \omega) v \wedge y - \theta(v \wedge x, \omega) u \wedge y \\ & + \theta(v \wedge y, \omega) u \wedge x - \theta(u \wedge y, \omega) v \wedge x) \end{aligned}$$

for $\omega, u \wedge v, x \wedge y \in V$. Then is again $\beta(\overline{\omega}, \mu) \in \mathfrak{gl}(n,\mathbb{C}) \subset \mathfrak{gl}(\wedge^2\mathbb{C}^n)$ and $Q_{\theta(.,\omega)}$ is also in the first prolongation because of the symmetry in the definition.

(ii) Weak-Berger: We show that the elements $Q_{\theta(.,u\wedge v)}(r \wedge s)$ generate the whole $\mathfrak{gl}(n,\mathbb{C})$. A direct calculation with the above θ-orthonormal base $e_i \wedge e_j$ and D_{ij} then gives

$$Q_{\theta(.,e_i\wedge e_j)}(e_k \wedge e_l) = \delta_{ik} D_{jl} - \delta_{jk} D_{il} - \delta_{il} D_{jk} + \delta_{jl} D_{ik}. \tag{C.2}$$

So we get that

$$Q_{\theta(.,e_i\wedge e_k)}(e_j \wedge e_k) \ = \ D_{ij} + \delta_{ij} D_{kk} \quad \text{for } i \neq k \neq j.$$

which shows that the whole $\mathfrak{gl}(n,\mathbb{C})$ is generated by $\beta(\overline{\omega}, \mu)$ since $n > 2$. So we have $\mathfrak{gl}(n,\mathbb{C}) \subset \widetilde{\mathfrak{gl}(n,\mathbb{C})} \subset \mathfrak{gl}(\wedge^2\mathbb{C}^n)$.

(iii) Berger: Again β satisfies the relation (4.60):

$$\begin{aligned} \overline{Q_{\theta(.,e_i\wedge e_j)}(e_k \wedge e_l)} \ &= \ -\delta_{ik} D_{lj} + \delta_{jk} D_{li} + \delta_{il} D_{kj} - \delta_{jl} D_{ki} \\ &= \ -Q_{\theta(.,e_k\wedge e_l)}(e_i \wedge e_j). \end{aligned}$$

(iv) The Riemannian symmetric space: We consider the compact, Kählerian, Riemannian symmetric space of type $D\ III$ given by the quotient $SO(2n,\mathbb{R})/U(n)$. For the tangent space we have

$$\mathfrak{so}(2n,\mathbb{R})/\mathfrak{u}(n) \simeq \left\{ \left(\begin{array}{cc} A & B \\ B & -A \end{array} \right) \middle| A, B \in \mathfrak{so}(n,\mathbb{R}) \right\} \simeq \mathbb{R}^{n(n-1)}. \tag{C.3}$$

The holonomy representation is given by the adjoint representation of $\mathfrak{u}(n)$ and isomorphic to the reellification of the representation in the third entry of table 2 via the following isomorphism

$$\begin{aligned} (\wedge^2\mathbb{C}^n)_{\mathbb{R}} &\simeq \left\{ \left(\begin{array}{cc} A & B \\ B & -A \end{array} \right) \middle| A, B \in \mathfrak{so}(n,\mathbb{R}) \right\} \\ e_i \wedge e_j &\mapsto \left(\begin{array}{cc} E_{ij} & 0 \\ 0 & -E_{ij} \end{array} \right) \\ i\, e_i \wedge e_j &\mapsto \left(\begin{array}{cc} 0 & E_{ij} \\ E_{ij} & 0 \end{array} \right) \end{aligned} \tag{C.4}$$

where E_{ij} is the standard basis of $\mathfrak{so}(n)$.
All together gives that $\mathfrak{g}_0$ is weak-Berger, but Berger too and the holonomy algebra of the Riemannian symmetric space of type $D\ III$.

C.4 $\mathfrak{sl}(\mathfrak{gl}(n,\mathbb{C}) \oplus \mathfrak{gl}(m,\mathbb{C}))$ acting on $V := \mathbb{C}^n \otimes \mathbb{C}^m$

The compact real form is

$$\mathfrak{s}(\mathfrak{u}(n) \oplus \mathfrak{u}(m)) := \mathfrak{u}(n) \oplus \mathfrak{u}(m) \cap \mathfrak{su}(n+m),$$

and the representation on $\mathbb{C}^n \otimes \mathbb{C}^m$ is of type 2.

$$\begin{array}{ccccc} \mathfrak{s}(\mathfrak{u}(n) \oplus \mathfrak{u}(m)) & \subset & \left\{ \begin{array}{c} \mathfrak{su}(n+m) \\ \mathfrak{sl}(\mathfrak{gl}(n,\mathbb{C}) \oplus \mathfrak{gl}(m,\mathbb{C})) \end{array} \right\} & \subset & \mathfrak{sl}(n+m,\mathbb{C}) \\ \downarrow & & \downarrow & & \\ \mathfrak{u}(n \cdot m) & \subset & \mathfrak{gl}(n \cdot m, \mathbb{C}) & & . \end{array}$$

For the hermitian form holds that

$$\theta(x \otimes u, y \otimes v) = \theta_n(x,y)\theta_m(u,v)$$

for $x, y \in \mathbb{C}^n$ and $u, v \in \mathbb{C}^m$. So if we have an θ_n-orthonormal base $(e_i, i = 1, \dots n)$ on $\mathbb{C}^n$ and a θ_m orthonormal $(e_p, p = 1, \dots m)$, then $e_i \otimes e_p$ is an θ-orthonormal base of $\mathbb{C}^n \otimes \mathbb{C}^m$.
The standard basis (D_{ij}, D_{pq}) of $\mathfrak{sl}(\mathfrak{gl}(n,\mathbb{C}) \oplus \mathfrak{gl}(m,\mathbb{C}))$ acts as follows

$$(D_{ij}, D_{pq}) \;:\; e_k \otimes e_r \longmapsto D_{ij}e_k \otimes e_r - e_k \otimes D_{pq}e_q = \delta_{ik}e_j \otimes e_r - \delta_{pr}e_k \otimes e_q.$$

Again we check the three points.

(i) The isomorphism between $\overline{V} \overset{\beta}{\simeq} \mathfrak{sl}\left(\mathfrak{gl}(n,\mathbb{C}) \oplus \mathfrak{gl}(m,\mathbb{C})\right)^{(1)}$ is defined by

$$\begin{aligned} &\mathfrak{gl}(n \cdot m, \mathbb{C}) \supset \mathfrak{gl}(n,\mathbb{C}) \oplus \mathfrak{gl}(m,\mathbb{C}) \ni \beta(\overline{\tau}, x \otimes u) := Q_{\theta(.,\tau)}(x \otimes u) \;: \\ &\qquad\qquad y \otimes v \longmapsto \theta(y \otimes u, \tau)x \otimes v + \theta(x \otimes v, \tau)y \otimes u \end{aligned}$$

for $\tau, x \otimes u, y \otimes v \in V$. Then is again $\beta(\overline{\tau}, \eta) \in \mathfrak{sl}(\mathfrak{gl}(n,\mathbb{C}) \otimes \mathfrak{gl}(m,\mathbb{C})) \subset \mathfrak{gl}(n \cdot m, \mathbb{C})$ and $Q_{\theta(.,\tau)}$ is also in the first prolongation because of the symmetry in the definition.

(ii) Weak-Berger: A direct calculation with the above θ- orthonormal base $e_i \otimes e_p$ and (D_{ij}, D_{pq}) then gives

$$Q_{\theta(.,e_i\otimes e_p)}(e_j \otimes e_q) = (\delta_{pq}D_{ij}, -\delta_{ij}D_{pq}) \tag{C.5}$$

which generates $\mathfrak{sl}(\mathfrak{gl}(n,\mathbb{C}) \otimes \mathfrak{gl}(m,\mathbb{C}))$. So the algebra is weak-Berger.

(iii) Berger: Again β satisfies the relation (4.60):

$$\begin{aligned} \overline{Q_{\theta(.,e_i\otimes e_p)}(e_j \otimes e_q)} &= (-\delta_{pq}D_{ji}, \delta_{ij}D_{qp}) \\ &= -Q_{\theta(.,e_j\otimes e_q)}(e_i \otimes e_p). \end{aligned}$$

(iv) The Riemannian symmetric space: We consider the compact, Kählerian, Riemannian symmetric space of type $A\ III$ given by the quotient $SU(n+m)/U(n)\cdot U(m)$. For tangent space we have

$$\mathfrak{su}(n+m)/\mathfrak{s}(\mathfrak{u}(n)\oplus\mathfrak{u}(m)) \simeq Mat_{\mathbb{C}}(n,m)_{\mathbb{R}} \simeq (\mathbb{C}^n \otimes \mathbb{C}^m)_{\mathbb{R}}\,.$$

This isomorphism identifies the holonomy representation and the representation in table 2. All together gives that $\mathfrak{g}_0$ is weak-Berger, but Berger too and the holonomy algebra of the Riemannian symmetric space of type $A\ III$.

C.5 $\mathbb{C}\ Id \oplus \mathfrak{spin}(10,\mathbb{C})$ acting on $V := \mathbb{C}^{16}$ and $\mathbb{C}\ Id \oplus \mathfrak{e}_6$ acting on $\mathbb{C}^{27}$

For these algebras we will not verify in detail that they come from real Lie algebras which are Berger and weak-Berger. Here we only will give the Riemannian symmetric spaces, the holonomy representation of which corresponds to the last two representations in table two. This we will do in detail, so that we have to recall its definitions with the help of the exceptional Lie algebras. These we will define in terms of spin representations following completely [Ada96].

C.5.1 The Lie algebras and groups of type E_8, E_7 and E_6

Let S^+_{16} be the irreducible real spinor module of $Spin(16)$ and Δ^+_{16} the complex one. The latter is — in dimension 16 — equipped with a real structure, which entails that $(S^+_{16})^{\mathbb{C}} = \Delta^+_{16}$. Since $Spin(16)$ is compact, Δ^+_{16} is unitary, a fact which makes it possible to define a $Spin(16)$-invariant inner product $\langle .,. \rangle$ on Δ^+_{16} with the help of the real structure. Since $Spin(16,\mathbb{C})$ is simple, we have an invariant inner product $(.,.)$ on

$\mathfrak{spin}(16,\mathbb{C})$ defined by the Killing form. Then the Lie algebra $\mathfrak{e}_8$ is defined on the complex vector space

$$\begin{aligned} \mathfrak{e}_8 &= \mathfrak{spin}(16,\mathbb{C}) \oplus \Delta_{16}^+ \quad \text{with the commutator:} \\ [A,B] &= [A,B]_{\mathfrak{spin}(16,\mathbb{C})} \ , \text{ for } A,B \in \mathfrak{spin}(16,\mathbb{C}) \\ [A,v] &= A \cdot v \ , \text{ for } A \in \mathfrak{spin}(16,\mathbb{C}), v \in \Delta_{16}^+ \\ [u,v] &\in \mathfrak{spin}(16,\mathbb{C}) \text{ defined by } \langle [A,u],v\rangle = (A,[u,v]) \text{ for all } A \in \mathfrak{spin}(16,\mathbb{C}). \end{aligned}$$

The last commutator is a kind of transposition of the mapping $\mathfrak{spin}(16,\mathbb{C})\otimes\Delta_{16}^+ \longrightarrow \Delta_{16}^+$ using that the 1-form on $\mathfrak{spin}(16,\mathbb{C})$ mapping A to $\langle [A,u],v\rangle$ can be identified with an element $B_{u,v} \in \mathfrak{spin}(16,\mathbb{C})$ with the help of $(.,.)$ via $\langle [A,u],v\rangle = (A, B_{u,v})$.

To verify that the defined $[.,.]$ is a Lie bracket, one calculates using the invariance of $(.,.)$ and $\langle .,. \rangle$ that the equation

$$\begin{aligned} &(B,[A,[u,v]]) + (B,[u,[v,A]]) + (B,[v,[A,u]]) = \\ &(B,A[u,v]) - (B,[u,v]A) - (B,[u,Av]) + (B,[v,Au]) = \\ &\qquad - ([A,B],[u,v]) - \langle Bu, Av\rangle + \langle Bv, Au\rangle = \\ &\qquad\qquad - \langle [A,B]u, v\rangle + \langle ABu, v\rangle - \langle BAu, v\rangle = 0 \end{aligned}$$

holds for every $B \in \mathfrak{spin}(16,\mathbb{C})$. But this implies the Jacobi identity.

Then $(.,.) + \langle .,. \rangle$ is an invariant inner product on $\mathfrak{e}_8$ proportional to the Killing form of $\mathfrak{e}_8$, so it is simple.

To get the real forms of $\mathfrak{e}_8$ one performs this construction with the real Lie algebra $\mathfrak{spin}(16)$ and the real spinor module S_{16}^+. The first two commutators are defined in the same manner as in the complex case. For the third commutator there is freedom choosing the inner products positive or negative definite. Choosing both positive or negative definite leads to the compact real form of $\mathfrak{e}_8$ denoted by $\mathfrak{e}_{8(-248)}$.

The number -248 indicates the character of this real form, i.e. the number $dim\ \mathfrak{p} - dim\ \mathfrak{k}$, where $\mathfrak{p} \oplus \mathfrak{k}$ is the Cartan decomposition of a real form. For compact real forms it is equal to the dimension of the Lie algebra (see for example [Hel78], Ch. X, §6, Nr. 2). The other real forms are $\mathfrak{e}_{8(8)}$ and $\mathfrak{e}_{8(-24)}$, which are not compact.

In the following we will only consider the compact case. The group E_8 is defined as the identity component of the group of automorphisms of $\mathfrak{e}_{8(-248)}$. Then E_8 has the Lie algebra $\mathfrak{e}_{8(-248)}$, is compact and contains $Spin(16)$.

To define E_6 and E_7 we consider the following two inclusions

$$\begin{array}{ccccccccc} & & Spin(4) & \hookrightarrow & Spin(12)\times Spin(4) & \hookrightarrow & Spin(16) & \hookrightarrow & E_8 \\ & \nearrow & \downarrow & & & & & & \\ SU(2) & \hookrightarrow & SO(4) & & \text{and} & & & & \end{array}$$

$$\begin{array}{ccccccccc} & & Spin(6) & \hookrightarrow & Spin(10)\times Spin(6) & \hookrightarrow & Spin(16) & \hookrightarrow & E_8 \\ & \nearrow & \downarrow & & & & & & \\ SU(3) & \hookrightarrow & SO(6) & . & & & & & \end{array}$$

Then E_7 and E_6 are defined as

$$\begin{aligned} E_7 &= \text{identity component of the centralizer of } SU(2) \text{ in } E_8 \\ E_6 &= \text{identity component of the centralizer of } SU(3) \text{ in } E_8 \end{aligned}$$

Both are compact and have the compact Lie algebras $\mathfrak{e}_{7(-133)}$ and $\mathfrak{e}_{6(-78)}$. It holds

$$E_6 \subset E_7 \subset E_8.$$

C.5.2 Decomposition of $\mathfrak{e}_8$

Now we will consider $\mathfrak{e}_8$ as irreducible module of E_8 under the adjoint representation and decompose it into irreducible modules for certain subgroups of E_8, producing the open representations in table 2 and 3.

First we take $\mathfrak{e}_8$ as a $Spin(16)$ module. Clearly it decomposes as follows

$$\mathfrak{e}_8 = \mathfrak{spin}(16,\mathbb{C}) \oplus \Delta^+_{16}.$$

We continue and take it as a module of $Spin(10)\times Spin(6)$. Then it decomposes as follows

$$\mathfrak{e}_8 = (\mathfrak{spin}(10,\mathbb{C})\otimes 1) \oplus (1\otimes \mathfrak{spin}(6,\mathbb{C})) \oplus \left(\mathbb{C}^{10}\otimes\mathbb{C}^6\right) \oplus \left(\Delta^+_{10}\otimes\Delta^+_6\right) \oplus \left(\Delta^-_{10}\otimes\Delta^-_6\right),$$

where $\mathbb{C}^{10}$ and $\mathbb{C}^6$ are the vector representations and Δ^+_{10}, Δ^-_{10}, Δ^+_6 and Δ^-_6 are the positive and negative complex spinor representations of of $Spin(10)$ and $Spin(6)$.

In the next step we take the group $Spin(10)\times U(3) \subset Spin(10)\times Spin(6)$. Then the summands decomposes as follows:

- $\mathfrak{spin}(6,\mathbb{C}) \simeq \mathfrak{so}(6,\mathbb{C})$ decomposes in $\mathfrak{gl}(3,\mathbb{C}) \oplus \left(\wedge^2\mathbb{C}^3 \oplus \wedge^2\overline{\mathbb{C}}^3\right)$. This is obvious if one complexifies the relation (C.3) and (C.4), taking into account that $(V_{\mathbb{R}})^{\mathbb{C}} = V\oplus\overline{V}$.

- The relation $\mathbb{C}^6 = (\mathbb{R}^6)^{\mathbb{C}} = \left(\mathbb{C}^3_{\mathbb{R}}\right)^{\mathbb{C}} = \mathbb{C}^3 \oplus \overline{\mathbb{C}}^3$ gives the decomposition of $\mathbb{C}^6$.

If one now writes $U(3) = SO(2)\times SU(3) = S^1 \times SU(3)$ one gets a further decomposition:

- $\mathfrak{spin}(6,\mathbb{C}) = \mathbb{C}\otimes 1 \oplus 1\otimes \mathfrak{sl}(3,\mathbb{C}) \oplus \left(\mathbb{C}\otimes \wedge^2\mathbb{C}^3 \oplus \overline{\mathbb{C}}\otimes \wedge^2\overline{\mathbb{C}}^3\right)$. Now from representation theory of $SU(3)$ follows that the module $\wedge^2\mathbb{C}^3$ is conjugate to $\mathbb{C}^3$, i.e. $\wedge^2\overline{\mathbb{C}}^3 \simeq \mathbb{C}^3$. So we get

$$\mathfrak{spin}(6,\mathbb{C}) = \mathbb{C}\otimes 1 \oplus 1\otimes \mathfrak{sl}(3,\mathbb{C}) \oplus \left(\mathbb{C}\otimes \overline{\mathbb{C}}^3 \oplus \overline{\mathbb{C}}\otimes \mathbb{C}^3\right).$$

- Δ_6^+ decomposes in the one-dimensional space which is fixed by $SU(3)$ and the remaining 3-dimensional space. In detail: Since for the volume element $e_1\cdot\ldots\cdot e_6$ in $Spin(6)$ holds that $(e_1\cdot\ldots\cdot e_6)^2 = (-1)^3 = -1$ it corresponds to $i\in S^1$. Therefore $i = e_1\cdot\ldots\cdot e_6$ acts on Δ_6^+ as multiplication with $i^3 = -i$. The remaining 3-dimensional space is isomorphic to $\mathbb{C}^3$. So Δ_6^+ decomposes as $S^1\times SU(3)$-module in $\overline{\mathbb{C}}\otimes 1 \oplus \overline{\mathbb{C}}\otimes \mathbb{C}^3$. Since $\Delta_6^- = \overline{\Delta_6^+}$ the decomposition of Δ_6^- is $\mathbb{C}\otimes 1\oplus \mathbb{C}\otimes\overline{\mathbb{C}}^3$.

All together we can write $\mathfrak{e}_8$ as $Spin(10)\times S^1\times SU(3)$- module as follows

$$\begin{aligned}
\mathfrak{e}_8 &= (\mathfrak{spin}(10,\mathbb{C})\otimes 1\otimes 1)\oplus(1\otimes\mathbb{C}\otimes 1)\oplus(1\otimes 1\otimes\mathfrak{sl}(3,\mathbb{C}))\\
&\quad \oplus 1\otimes\left(\mathbb{C}\otimes\overline{\mathbb{C}}^3\oplus\overline{\mathbb{C}}\otimes\mathbb{C}^3\right)\oplus\mathbb{C}^{10}\otimes\left(\mathbb{C}\otimes\mathbb{C}^3\oplus\mathbb{C}\otimes\overline{\mathbb{C}}^3\right)\\
&\quad \oplus\Delta_{10}^+\otimes\left(\overline{\mathbb{C}}\otimes 1\oplus\overline{\mathbb{C}}\otimes\mathbb{C}^3\right)\oplus\Delta_{10}^-\otimes\left(\mathbb{C}\otimes 1\oplus\mathbb{C}\otimes\overline{\mathbb{C}}^3\right)\\
&= \left[\,(\mathfrak{spin}(10,\mathbb{C})\otimes 1)\oplus(1\otimes\mathbb{C})\oplus(\Delta_{10}^+\otimes\overline{\mathbb{C}}\oplus\Delta_{10}^-\otimes\mathbb{C})\,\right]\otimes 1\\
&\quad \oplus 1\otimes 1\otimes\mathfrak{sl}(3,\mathbb{C})\\
&\quad \oplus\left[\,(1\otimes\overline{\mathbb{C}})\oplus(\mathbb{C}^{10}\otimes\mathbb{C})\oplus(\Delta_{10}^+\otimes\overline{\mathbb{C}})\,\right]\otimes\mathbb{C}^3\\
&\quad \oplus\left[\,(1\otimes\mathbb{C})\oplus\left(\overline{\mathbb{C}}^{10}\otimes\mathbb{C}\right)\oplus(\Delta_{10}^-\otimes\mathbb{C})\,\right]\otimes\overline{\mathbb{C}}^3
\end{aligned}$$

This decomposition has two important consequences. The first is a description of the complexified Lie algebra of E_6. Since E_6 is the centralizer of $SU(3)$ in E_8 we have that

$$\mathfrak{e}_6 = \mathfrak{spin}(10,\mathbb{C})\oplus\mathbb{C}\oplus\left(\Delta_{10}^+\otimes\overline{\mathbb{C}}\oplus\Delta_{10}^-\otimes\mathbb{C}\right).$$

This implies that $\mathfrak{spin}(10,\mathbb{C})\oplus\mathbb{C}\subset\mathfrak{e}_6$ with

$$\mathfrak{e}_6/\left(\mathfrak{spin}(10,\mathbb{C})\oplus\mathbb{C}\right) = \Delta_{10}^+\otimes\overline{\mathbb{C}}\oplus\Delta_{10}^-\otimes\mathbb{C}. \tag{C.6}$$

for the compact real form this means that

$$\mathfrak{spin}(10)\oplus\mathfrak{so}(2)\subset\mathfrak{e}_{6(-78)}. \tag{C.7}$$

The second is that we can write $\mathfrak{e}_8$ as a $E_6\times SU(3)\subset E_8$ module as follows

$$\mathfrak{e}_8 = \mathfrak{e}_6\oplus\mathfrak{sl}(3,\mathbb{C})\oplus\underbrace{\left[\overline{\mathbb{C}}\oplus\mathbb{C}^{10}\oplus\Delta_{10}^+\right]}_{:=V}\otimes\mathbb{C}^3\oplus\underbrace{\left[\mathbb{C}\oplus\overline{\mathbb{C}}^{10}\oplus\Delta_{10}^-\right]}_{:=\overline{V}}\otimes\overline{\mathbb{C}}^3.$$

This gives that V and $\overline{V}$ are irreducible, 27-dimensional E_6 modules which are conjugate to each other since $\overline{\Delta_{10}^+} \simeq \Delta_{10}^-$ and $\mathbb{C}^{10}$ is self conjugate as $Spin(10)$-module.
Now one can clarify the last open representation of table 2. It is the representation of $\mathbb{C} \oplus \mathfrak{e}_6$ on $V \simeq \mathbb{C}^{27}$.
These facts enables us to describe the remaining symmetric spaces.

C.5.3 The exceptional Kählerian symmetric spaces

The space $(\mathfrak{e}_{6(-78)}, \mathfrak{so}(2) \oplus \mathfrak{spin}(10))$ The inclusion $\mathfrak{so}(2) \oplus \mathfrak{spin}(10) \subset \mathfrak{e}_{6(-78)}$ was described in the previous section. The complexified holonomy representation is

$$\left(\mathfrak{e}_{6(-78)}/\mathfrak{so}(2) \oplus \mathfrak{spin}(19)\right)^{\mathbb{C}} = \mathfrak{e}_6/\mathbb{C} \oplus \mathfrak{spin}(10, \mathbb{C}) = \Delta_{10}^+ \otimes \overline{\mathbb{C}} \oplus \Delta_{10}^- \otimes \mathbb{C}.$$

Thats why the holonomy representation is $\left(\Delta_{10}^+ \otimes \overline{\mathbb{C}}\right)_{\mathbb{R}}$ which is the reellification of the representation number 5 in table 2. But this was the result to show.
To describe the symmetric space completely one has to find the discrete kernel of the representation of $Spin(10) \times S^1$ on $\Delta_{10}^+ \otimes \overline{\mathbb{C}} \oplus \Delta_{10}^- \otimes \mathbb{C}$. But this kernel is equal to $\mathbb{Z}_4$ generated by $(e_1 \cdot \ldots \cdot e_{10}, i)$, since $e_1 \cdot \ldots \cdot e_{10}$ acts as $i^5 = i$ on Δ_{10}^+ such that $e_1 \cdot \ldots \cdot e_{10} \times i$ acts as $i \cdot (-i) = 1$ on $\Delta_{10}^+ \otimes \overline{\mathbb{C}}$.
So the exceptional Kählerian symmetric space corresponding to $(\mathfrak{e}_{6(-78)}, \mathfrak{so}(2) \oplus \mathfrak{spin}(10)$ is the following quotient

$$E_6 \Big/ {S^1 \times Spin(10)/\mathbb{Z}_4} .$$

The space $(\mathfrak{e}_{7(-133)}, \mathfrak{so}(2) \oplus \mathfrak{e}_{6(-78)})$ The holonomy representation of this space is 54-dimensional and irreducible of type two. I.e. it is the reellification of one of the 27-dimensional complex representations of E_6. Since they are conjugate to each other it does not depend which one we take. So we have proved proposition 4.59 also for the last case.

Bibliography

[Ada96] John Frank Adams. *Lectures on Exceptional Lie Groups*. Chicago Lectures in Mathematics. University of Chicago Press, 1996.

[Ale68] Dimitri Alekseevsky. Riemannian spaces with unusual holonomy groups. *Funct. Anal. Appl.*, 2:97 – 105, 1968.

[AS53] W. Ambrose and I. M. Singer. A theorem on holonomy. *Amer. Math. Soc.*, 79:428–443, 1953.

[Bär93] Christian Bär. Real Killing spinors and holonomy. *Commun. Math. Phys.*, 154(3):509–521, 1993.

[Bau81] Helga Baum. *Spin-Strukturen und Dirac-Operatoren über pseudoriemannschen Mannigfaltigkeiten*, volume 41 of *Teubner-Texte zur Mathematik*. Teubner-Verlagsgesellschaft, 1981.

[Bau94] Helga Baum. A remark on the spectrum of the Dirac operator on pseudo-Riemannian spin maniofolds. SFB 288-Preprint, nr.136, 1994.

[Bau99a] Helga Baum. Lorentzian twistor spinors and CR-geometry. *Differ. Geom. Appl.*, 11(1):69–96, 1999.

[Bau99b] Helga Baum. Twistor spinors on Lorentzian manifolds, CR-geometry and Fefferman spaces. In *Kolr, Ivan (ed.) et al., Differential geometry and applications. Proceedings of the 7th international conference, DGA 98, and satellite conference of ICM in Berlin, Brno, Czech Republic, August 10-14, 1998. Brno: Masaryk University. 29-37* . 1999.

[Bau00a] Helga Baum. Twistor and Killing spinors in Lorentzian geometry. In *Bourguignon, Jean Pierre (ed.) et al., Global analysis and harmonic analysis. Papers from the conference, Marseille-Luminy, France, May 1999. Paris: Socit Mathmatique de France. Smin. Congr. 4, 35-52* . 2000.

[Bau00b] Helga Baum. Twistor spinors on Lorentzian symmetric spaces. *J. Geom. Phys.*, 34(3-4):270–286, 2000.

[Bau03] Helga Baum. Conformal Killing spinors and special geometric structures in Lorentzian geometry - a survey, 2003.

[Ber55] Marcel M. Berger. Sur les groupes d'holonomie homogène des variétés a connexion affine et des variétés riemanniennes. *Bull. Soc. Math. France*, 83:279–330, 1955.

[Ber57] Marcel M. Berger. Les espace symétriques non compacts. *Ann. Sci. École Norm. Sup.*, 74:85–177, 1957.

[Bes87] Arthur L. Besse. *Einstein Manifolds.* Springer Verlag, Berlin-Heidelberg-New York, 1987.

[BFOHP02] Matthias Blau, Jose Figueroa-O'Farrill, Christopher Hull, and George Papadopoulos. Penrose limits and maximal supersymmetry. *Class.Quant.Grav.*, 19:L87–L95, 2002.

[BG72] R. Brown and Alfred Gray. Riemannian manifolds with holonomy group $Spin(9)$. In *Differential geometry (in honor of Kentaro Yano)*, pages 41 – 59. Kinokuniya, Tokyo, 1972.

[BI93] Lionel Berard-Bergery and Aziz Ikemakhen. On the holonomy of Lorentzian manifolds. *Proceedings of Symposia in Pure Mathematics*, 54(2):27–40, 1993.

[BI97] Lionel Berard Bergery and Aziz Ikemakhen. Sur l'holonomie des variétés pseudo-Riemanniennes de signature (n,n). *Bull. Soc. math. France*, 125:93–114, 1997.

[BK99] Helga Baum and Ines Kath. Parallel spinors and holonomy groups on pseudo-Riemannian spin manifolds. *Annals of Global Analysis and Geometry*, 17:1–17, 1999.

[BK03] Helga Baum and Ines Kath. Doubly Extended Lie Groups - Curvature, Holonomy and Parallel Spinors, 2003. arXiv:math.DG/0203189.

[BL52] G. Borel and A. Lichnerowicz. Groupes d'holonomie des variétés riemanniennes. *Acad. Sci. Paris*, 234:1835–1837, 1952.

[BL03] Helga Baum and Felipe Leitner. The twistor equation in Lorentzian spin geometry, 2003. arXiv:math.DG/0305063.

[Boh03] Christoph Bohle. Killing spinors on Lorentzian manifolds. *J. Geom. Phys.*, 45(3-4):285–308, 2003.

[Bou75] Nicholas Bourbaki. *Groupes et algèbres de Lie*, volume 38 of *Éléments de matématiques*, chapter 8. Hermann, 1975.

[Bou82] Nicholas Bourbaki. *Groupes et algèbres de Lie*, chapter 9. Éléments de matématiques. Masson, 1982.

[Bou00] Charles Boubel. *Sur l'holonomie des variétés pseudo-riemanniennes.* PhD thesis, Université Henri Poincaré, Nancy, 2000.

[Bri25] H. W. Brinkmann. Einstein spaces which are mapped conformally on each other. *Math. Ann.*, 94:119–145, 1925.

[Bry87] Robert L. Bryant. Metrics with exceptional holonomy. *Annals of Mathematics*, 126(2):525–576, 1987.

[Bry99a] Robert L. Bryant. Recent advances in the theory of holonomy. *Séminair BOURBAKI*, 51(861):1–24, 1999.

[Bry99b] Robert L. Bryant. Remarks on spinors in low dimensions. Unpublished note, 1999.

[Bry99c] Robert L. Bryant. Spin(10,1)-metrics with a parallel null spinor and maximal holonomy. Unpublished note, 1999.

[Bry00] Robert L. Bryant. Pseudo-Riemannian metrics with parallel spinor fields and vanishing Ricci tensor. *Global Analysis and Harmonic Analysis, Séminaires et Congrès*, 4:53–93, 2000.

[BS89] Robert L. Bryant and Simon M. Salamon. On the construction of some complete metrics with exceptional holonomy. *Duke Math. J.*, 58(3):829–580, 1989.

[BZ03] Charles Boubel and Abdelghani Zeghib. Dynamics of some Lie subgroups of $O(n,1)$, applications, 2003. Preprint, `http://www.umpa.ens-lyon.fr/~zeghib/pubs.html`.

[Car09] Elie Cartan. Les groupes de transformations continus, infinis, simples. *Ann. Ec. Norm.*, 26:93–161, 1909.

[Car14] Elie Cartan. Les groupes projectifs continus réels qui ne laissant invariante aucune multiplicité plane. *Journ. Math. pures et appl.*, 10:149–186, 1914. or Œvres complètes, vol. 1, pp. 493-530.

[Car23a] Elie Cartan. Sur variétés à connexion affine et la théorie de la relativité généralisée i. *Ann. Sci. Ec. Norm. Sup.*, 40:325–412, 1923. or Œvres complètes, tome III, 659-746.

[Car23b] Elie Cartan. Sur variétés à connexion affine et la théorie de la relativité généralisée ii. *Ann. Sci. Ec. Norm. Sup.*, 41:1–25, 1923. or Œvres complètes, tome III, 799-824.

[Car26] Elie Cartan. Les groupes d'holonomie des espaces généralisés. *Acta Math.*, 48:1–42, 1926. or Œvres complètes, tome III, vol.2, 997-1038.

[CMS95] Quo-Shin Chi, Sergey A. Merkulov, and Lorenz J. Schwachhöfer. On the Incompleteness of Berger's List of Holonomy Representations, 1995.

[CP80] M. Cahen and M. Parker. *Pseudo-Riemannian Symmetric Spaces*, volume 24, No.229 of *Memoirs of the AMS*. American Mathematical Society, 1980.

[CW70] M. Cahen and N. Wallach. Lorentzian symmetric spaces. *Bull. Amer. Math. Soc.*, 79:585–591, 1970.

[DNP86] M. Duff, B. Nilsson, and C. Pope. Kaluza-Klein supergravity. *Phys. Rep.*, 130:1–142, 1986.

[dR52] Georges de Rham. Sur la réducibilité d'un espace de Riemann. *Math. Helv.*, 26:328–344, 1952.

[dSO01] Antonio J. di Scala and Carlos Olmos. The geometry of homogeneous submanifolds in hyperbolic space. *Mathematische Zeitschrift*, 237(1):199–209, 2001.

[Eis38] Luther Pfahler Eisenhardt. Fields of parallel vectors in Riemannian space. *Annals of Mathematics*, 39:316–321, 1938.

[FO00] José Miguel Figueroa-O'Farrill. Breaking the M-waves. *Classical Quantum Gravity*, 17(15):2925–2947, 2000.

[FO02] José Figueroa-O'Farrill. On parallelisable NS-NS backgrounds. EMPG-03-09, 2002. arXiv:hep-th/0305079.

[FOP02] José Figueroa-O'Farrill and George Papadopoulos. Maximally supersymmetric solutions of ten- and eleven-dimensional supergravities. EMPG-02-16, 2002. arXiv:hep-th/0211089.

[Fri80] Th. Friedrich. Der erste Eigenwert des Dirac-Operators einer kompakten, Riemannschen Mannigfaltigkeit nichtnegativer Skalarkrmmung. *Math. Nachr.*, 97:117–146, 1980.

[Gal03] Anton S. Galaev. The spaces of curvature tensors for holonomy algebras of Lorentzian manifolds, 2003. arXiv:math.DG/0304407.

[Got78] Morikuni Goto. *Semisimple Lie Algebras*, volume 38 of *Lecture Notes in Pure and Applied Mathematics*. Marcel Dekker Inc., New York, Basel, 1978.

[Hal93] Graham S. Hall. Space-times and holonomy groups. In *Differential Geometry and Its Applications. Proc. Conf. Opava August 24-28, 1993*, pages 201–210. Silesian University, 1993.

[Hel78] Sigurdur Helgason. *Differential Geometry, Lie Groups, and Symmetric Spaces*. Academic Press, 1978.

[Hit74] Nigel Hitchin. Harmonic spinors. *Advances in Mathematics*, 14:1–55, 1974.

[HL00] Graham S. Hall and D. P. Lonie. Holonomy goups and space-times. *Class. Quantum Grav.*, 17:1369–1382, 2000.

[HO65] J. Hano and H. Ozeki. On the holonomy groups of linear connections. *Nagoya Math. Journal*, 10:97–100, 1965.

[Ike96] Aziz Ikemakhen. Examples of indecomposable non-irreducible Lorentzian manifolds. *Ann. Sci. Math. Québec*, 20(1):53–66, 1996.

[Ike99] Aziz Ikemakhen. Sur l'holonomie des variétés pseudo-Riemanniennes de signature (2,n+2). *Publicacions Mathemàtiques*, 43:3–29, 1999.

[Iwa59] Nagayoshi Iwahori. On real irreducible representations of Lie algebras. *Nagoya Mathematical Journal*, pages 59–83, 1959.

[Joy00] Dominic D. Joyce. *Compact Manifolds with Special Holonomy.* Oxford Math. Monographs. Oxford Univerity Press, 2000.

[Kat99] Ines Kath. *Killing Spinors on Pseudo-Riemannian Manifolds.* 1999. Habilitationsschrift, Humboldt-Universität Berlin.

[Kat00] Ines Kath. Parallel pure spinors on pseudo-Riemannian manifolds. In W.H. Chen, C.P. Wang, A.-M. Li, U. Simon, M. Wiehe, and L. Verstraelen, editors, *Differential Geometry in Honor of Prof. S.S.Chern*, volume X of *Geometry and Topology of Submanifolds*, pages 87–104. World Scientific, 2000. Peking University, China, 29.08-03.09.1999, TU Berlin, 26.-28.11.1999.

[KN63] Shoshichi Kobayashi and Katsumi Nomizu. *Foundations of Differential Geometry*, volume 1. Interscience Wiley, New York, 1963.

[KN65] Shoshichi Kobayashi and Tadashi Nagano. On filtered Lie algebras and geoemtric structures II. *J. Math. Mech.*, 14:513–521, 1965.

[Kna02] Anthony W. Knapp. *Lie Groups Beyond an Introduction*, volume 140 of *Progress in Mathematics*. Birkhäuser, 2 edition, 2002.

[Lei01] Felipe Leitner. *The Twistor equation in Lorentzian spin geometry.* PhD thesis, Humboldt University Berlin, 2001.

[Lei03] Felipe Leitner. Imaginary Killing Spinors in Lorenztian Geometry, 2003. arXiv:math.DG/0302024.

[LM89] H. Blaine Lawson and Marie-Louise Michelsohn. *Spin Geometry.* Princeton University Press, 1989.

[MS99] Sergei Merkulov and Lorenz Schwachhöfer. Classification of irreducible holonomies of torsion-free affine connections. *Ann. Math.*, 150:77–149, 1999.

[Neu02] Thomas Neukirchner. Pseudo-Riemannian symmetric spaces. Diplomarbeit, Humboldt-Universität zu Berlin, 2002.

[Nij52] Albert Nijenhuis. *Theory of the Geometric Object.* 1952.

[Nij53] Albert Nijenhuis. On the holonomy groups of linear connections. *Indagationes Matematicae*, 15:233–249, 1953.

[NW84] P. Nieuwenhuizen and N. P. Warner. Integrability conditions for Killing spinors. *Comm. Math. Phys.*, 93:277–284, 1984.

[OV94] Arkadij. L. Onishchik and E. B. Vinberg, editors. *Lie Groups and Lie Algebras III*, volume 41 of *Encyclopedia of Mathematical Sciences.* Springer, 1994.

[Sal89] Simon M. Salamon. *Riemannian Geometry and Holonomy Groups*, volume 201 of *Pitmann Research Lecture Notes.* 1989.

[Sch60] J. F. Schell. Classification of 4-dimensional Riemannian spaces. *J. of Math. Physics*, 2:202–206, 1960.

[Sch74] Rainer Schimming. Riemannsche Räume mit ebenfrontiger und mit ebener Symmetrie. *Mathematische Nachrichten*, 59:128–162, 1974.

[Sch99] Lorenz Schwachhöfer. *On the Classification of Holonomy Representations.* 1999. Habilitationsschrift, Mathematisches Institut der Universität Leipzig.

[Ser87] Jean-Pierre Serre. *Complex Semisimple Lie Algebras.* Springer, 1987.

[Sha70] Ronald Shaw. The subgroup structure of the homogeneous Lorentz group. *Quart. J. Math. Oxford*, 21:101–124, 1970.

[Sim62] James Simons. On the transitivity of holonomy systems. *Annals of Mathematics*, 76(2):213–234, September 1962.

[Tit67] Jaques Tits. *Tabellen zu den einfachen Lie-Gruppen*, volume 40 of *Lecture Notes in Math.* Springer-Verlag, Berlin, 1967.

[Wal49] A. G. Walker. On parallel fields of partially null vector spaces. *Quart. Journ. of Mathematics*, 20:135–145, September 1949.

[Wan89] McKenzie Y. Wang. Parallel spinors and parallel forms. *Ann. Global Anal. Geom.*, 7(1):59–68, 1989.

[Wu64] H. Wu. On the de Rham decomposition theorem. *Illinois J. Math.*, 8:291–311, 1964.

[Zeg03] Abdelghani Zeghib. Remarks on Lorentz symmetric spaces, 2003. Preprint, `http://www.umpa.ens-lyon.fr/~zeghib/pubs.html`.